TOPICS IN COMMUTATIVE RING THEORY

TOPICS IN COMMUTATIVE RING THEORY

JOHN J. WATKINS

PRINCETON UNIVERSITY PRESS
Princeton and Oxford

© 2007 by Princeton University Press

Published by Princeton University Press, 41 William Street, Princeton, New Jersey 08540

In the United Kingdom: Princeton University Press, 3 Market Place, Woodstock, Oxfordshire OX20 1SY

Library of Congress Cataloging-in-Publication Data

Watkins, John J.
Topics in commutative ring theory / John J. Watkins.
p. cm.
Includes bibliographical references and index.
ISBN-13: 978-0-691-12748-4 (acid-free paper)
ISBN-10: 0-691-12748-4 (acid-free paper)
1. Commutative rings. 2. Rings (Algebra) I. Title.

QA251.3W38 2007
512.44–dc22   2006052875

British Library Cataloging-in-Publication Data is available

This book has been composed in ITC Stone Serif

Printed on acid-free paper. ∞

pup.princeton.edu

Printed in the United States of America

10 9 8 7 6 5 4 3 2 1

For Jim Brewer

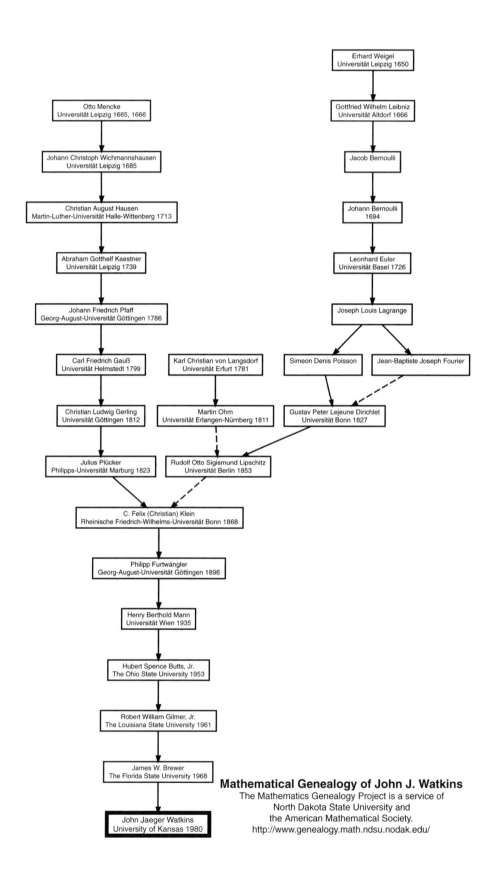

Mathematical Genealogy of John J. Watkins
The Mathematics Genealogy Project is a service of
North Dakota State University and
the American Mathematical Society.
http://www.genealogy.math.ndsu.nodak.edu/

Contents

Preface

This collection of lectures in commutative ring theory has grown out of a course I have taught for many years at Colorado College for advanced undergraduates taking a second course in abstract algebra and is intended as an introduction to abstract mathematics. It is abstraction — more than anything else — that characterizes the mathematics of the twentieth century. There is both power and elegance in the axiomatic method, attributes that can and should be appreciated by students early in their mathematical careers and even if they happen to be confronting contemporary abstract mathematics in a serious way for the very first time.

Commutative ring theory arose more than a century ago to treat age-old questions in geometry and number theory; it is is therefore, in part, a branch of applied mathematics in the sense that it is applied to other areas of mathematics. Even today it draws nourishment from these two subjects. But commutative ring theory is also very much a part of pure mathematics, and as such it has a life of its own that is quite independent of its origins. It is largely the balance — even the tension — between these two aspects of its personality that makes commutative ring theory such a rich and beautiful subject for study.

While many readers of this book may well have previously studied modern algebra, I will assume no particular knowledge on the part of the reader other than perhaps a passing awareness of what a group is. I believe that, by focusing our attention on a single relatively narrow field of modern abstract mathematics, we can begin at the beginning and take an enthusiastic reader on a trip far into the vibrant world of contemporary mathematics. The itinerary and pace for this journey have been conceived with advanced — and, I repeat, enthusiastic — undergraduates in mind, but I sincerely hope that graduate students beginning their specialization in algebra as well as seasoned mathematicians from other areas of mathematics will also find the journey worthwhile and pleasurable.

The intent, then, is a fairly leisurely and reader-friendly passage. Our goal is to get a feel for the lay of the land, to marvel at some of the vistas, and to poke around a few of the back roads. We may miss a couple of the main highways, and we will certainly resist climbing some of the higher peaks. We may not get as far, or as high, as some would like, but we prefer not to lose anyone along the way. There will always be time — and other guidebooks — for other, more ambitious, trips for those of you who have enjoyed this one.

I have placed a series of problems at the end of each chapter in order to encourage active reading. These exercises provide you with a way of immediately reinforcing new concepts, as well as becoming adept at some of the fundamental techniques of commutative ring theory. I have provided solutions to some of these exercises at the end of the book so that you can compare your work with what could be considered "standard" solutions. Some of the problems are routine exercises designed to build technical skill or reinforce basic new ideas, and advanced readers may well wish to skip most of them. However, many of the problems are in fact extremely important mathematical results in their own right, and will be used freely later on in the text.

I would like to thank first of all the many students at Colorado College who have been subjected to earlier, considerably rougher, versions of this book. In particular, I appreciate the enthusiastic and generous help I have received through the years from David Carlson, T.J. Calvert, Lisa Converse, Courtney Gibbons, Laura Hegerle, Eric Raarup, Karin Reisbeck, Chantelle Szczech-Jones, Mark Sweet, Rahbar Virk, and Trevor Wilson. I would also like to thank two colleagues, Doug Costa and Wojciech Kosek, whose suggestions have improved this book considerably. Mostly, though, I would like to thank Jim Brewer, my thesis advisor at the University of Kansas, from whom I not only first learned this beautiful subject but also received so very much in terms of guidance, wisdom, and friendship.

John J. Watkins
Colorado Springs
May 31, 2006

TOPICS IN COMMUTATIVE RING THEORY

1

Rings and Subrings

The Notion of a Ring

In 1888 — when he was only 26 years old — David Hilbert stunned the mathematical world by solving the main outstanding problem in what was then called invariant theory. The question that Hilbert settled had become known as *Gordan's Problem*, for it was Paul Gordan who, 20 years earlier, had shown that *binary* forms have a finite basis. Gordan's proof was long and laboriously computational; there seemed little hope of extending it to ternary forms, and even less of going beyond. We will not take the time here to explore any of the details of Gordan's problem or even the nature of invariant theory (and you shouldn't be at all concerned if you don't have the foggiest idea what binary or ternary forms are or what a basis is), but Hilbert — in a single brilliant stroke — proved that there is in fact a *finite* basis for all invariants, *no matter how high the degree.*

The structure of Hilbert's proof is really quite simple and is worth looking at (again we will not worry at all about most of the details). First, Hilbert showed that if a ring R has a certain property P, then the ring of polynomials in a single variable x with coefficients from the ring R also has that same property P. (Today we would say that P is the property that any ideal is finitely generated, but that is getting well ahead of our story.) We will use the convenient notation $R[x]$ to represent this ring of polynomials in x with coefficients from R, and so we can summarize Hilbert's first step as

if a ring R has property P, then so does the ring $R[x]$.

Next, Hilbert wanted to show that the ring of polynomials in *two* variables x and y with coefficients from the ring R also has property P. We represent this ring of polynomials in two variables by $R[x, y]$. These polynomials are just like ordinary polynomials we are used to such as $x^2 + 4$ and $2y^2 - y + 3$, except that now we can have the two variables x and y mixed together in a single polynomial. An example of such a polynomial is

$$3x^2 + 4xy + 2y^2 + 7x - 5y + 12,$$

where in this case the coefficients are all integers. Hilbert saw that this ring of polynomials $R[x, y]$ in *two* variables x and y with coefficients from the ring R can be thought of as the ring of polynomials in a *single* variable y with coefficients from the ring of polynomials in x. For example, we can write the above polynomial in two ways, depending on how we choose to group the terms:

$$3x^2 + 4xy + 2y^2 + 7x - 5y + 12 = 2y^2 + (4x - 5)y + (3x^2 + 7x + 12).$$

On the left the polynomial is written as a polynomial in the ring $R[x, y]$, whereas on the right it is written as a polynomial in one variable y where the coefficients are themselves polynomials in x. Our notation for this latter ring is $(R[x])[y]$ — or more simply $R[x][y]$ — emphasizing the fact that the coefficients are now polynomials in the ring $R[x]$.

Using this simple idea, Hilbert concluded that the ring of polynomials $R[x, y]$ also has property P. His argument went like this: since the ring R has property P, so does the ring $R[x]$; but then since the ring $R[x]$ has property P, so does the ring $R[x][y]$; and, as we have just seen, this latter ring is really the same as the ring $R[x, y]$.

In this way, by adding one variable at a time, Hilbert showed that the polynomial ring in any finite number of variables has property P. For example, we could now conclude that the ring $R[x, y, z]$ has property P since this ring is the same as the ring $R[x, y][z]$ and we have just argued that $R[x, y]$ has property P. The key to Hilbert's argument, then, is to verify his very first step — namely, that if a ring R has property P, then so does the polynomial ring with coefficients from R.

Now Hilbert did this not by explicitly constructing a basis (as Gordan had done for the binary case), but rather — and this is the brilliant part of his proof — by showing that if there were no finite basis, then a contradiction arises. Therefore, there *must* be a finite basis after all! Nowadays, we are very comfortable with such a *proof by contradiction*, but Hilbert had used this technique in a new way: he had proved the *existence* of something without actually constructing it. This existence proof did not meet with universal favor in the mathematical climate of his day. In fact, Gordan — hardly an impartial observer — chose this time to issue one of the most memorable lines in all of mathematics: "Das is nicht Mathematik. Das ist Theologie." It was not until four years later, when Hilbert was able to use the existence of a finite basis to show how such a basis could actually be constructed, that Gordan conceded: "I have convinced myself that theology also has its advantages."

At the heart of Hilbert's proof — and the attendant controversy — lies the abstract notion of a ring, though it would be several years until

Hilbert would actually provide us with the term *ring* (or *Zahlring* — literally, number ring) which we now use today. The idea is that, for instance, although polynomials certainly differ in many obvious ways from integers, there are ways in which polynomials and integers are similar: for example, you can add or multiply integers and you can also add or multiply polynomials. It is the differences between integers and polynomials that most of us notice first, but Hilbert focused instead on their similarities. So, the idea behind the notion of a ring is that integers, rationals, reals, complex numbers, polynomials with complex coefficients, and continuous functions, as different as all of these systems may appear to us, all share certain characteristics. It is these shared underlying characteristics which provide the basis for the following unifying axioms and our definition of a ring, for it is the abstract notion of a ring that so elegantly captures the essence of what these familiar mathematical systems share in their behavior.

The Definition of a Ring

Before we actually define a ring, let us talk a bit about what a ring is. Quite simply it is a set of elements (typically a set of numbers of some kind, or perhaps a set consisting of a particular type of function) together with two operations on those elements called addition and multiplication. It is very important to think of a ring as a single object consisting of *both* the underlying set and the two operations, and *not* just as a set by itself. Furthermore, these operations will need to behave the way we expect them to behave. For example, if a and b are two elements in a ring, we expect $a + b$ and $b + a$ to be equal, or we expect $a + 0$ to equal a, or we expect $a(a + b)$ to equal $a^2 + ab$. We have these expectations no matter whether a and b are numbers, or polynomials, or matrices.

Let us look at some specific examples of rings. In each case, note that we present both a set and two operations on that set in order to describe the ring.

Example 1

Certainly the single most fundamental example of a ring is the ring based on the numbers $\ldots, -3, -2, -1, 0, 1, 2, 3, \ldots$. We will call this ring the *ring of integers* or, more simply, *the integers*, and we will denote this ring by **Z**. This letter for the integers may seem peculiar to you at first, but it comes from *Zahl*, the German word for number, and serves as a nice reminder to us of the history of the notion of a ring. The ring **Z** then consists of the set of integers together with the ordinary operations of addition and multiplication.

Example 2

The most natural ring to consider next is the ring based on the numbers that are fractions of integers, such as $\frac{1}{3}$ and $\frac{22}{7}$. Thus, we will consider the *rational numbers* or, more simply, *the rationals* as a ring with the ordinary operations of addition and multiplication of fractions. This ring is denoted by **Q**. (By the way, why do you suppose mathematicians long ago decided on this particular letter to represent the rationals?)

Example 3

Similarly the *real numbers* or, more simply, *the reals* together with the ordinary operations of addition and multiplication form a ring which we denote by **R**.

Example 4

The *complex numbers* with their ordinary operations of addition and multiplication form a ring which we denote by **C**. A complex number is a number of the form $a + bi$, where a and b are real numbers and $i^2 = -1$. The two operations of addition and multiplication in this ring are completely natural — for example,

$$(1 + 7i) + (2 - 3i) = 3 + 4i$$

and

$$(1 + 7i) \cdot (2 - 3i) = 2 - 21i^2 - 3i + 14i = 23 + 11i,$$

since $i^2 = -1$.

Somewhat more formally, the two operations of addition and multiplication of complex numbers can be defined as follows, where $a, b, c, d \in \mathbf{R}$:

$$(a + bi) + (c + di) = (a + c) + (b + d)i$$

and

$$(a + bi) \cdot (c + di) = (ac - bd) + (ad + bc)i.$$

Example 5

The *set of polynomials with integer coefficients* together with the ordinary operations of addition and multiplication of polynomials — that is, you add and multiply polynomials just as you did in high school — also form a ring. We denote this ring by $\mathbf{Z}[x]$. So, for example,

$$(1 + x + x^2) + (-2 + 3x - x^3) = -1 + 4x + x^2 - x^3$$

and

$$(1 + x + x^2) \cdot (-2 + 3x - x^3) = -2 + x + x^2 + 2x^3 - x^4 - x^5.$$

Example 6

The set of *2 by 2 matrices* whose entries are real numbers together with the ordinary operations of matrix addition and multiplication form a ring. We denote this ring by $\mathbf{M_2}$. Formally, the two operations for $\mathbf{M_2}$ are defined as follows, where $a, b, c, d, e, f, g, h \in \mathbf{R}$:

$$\begin{pmatrix} a & b \\ c & d \end{pmatrix} + \begin{pmatrix} e & f \\ g & h \end{pmatrix} = \begin{pmatrix} a+e & b+f \\ c+g & d+h \end{pmatrix}$$

and

$$\begin{pmatrix} a & b \\ c & d \end{pmatrix} \cdot \begin{pmatrix} e & f \\ g & h \end{pmatrix} = \begin{pmatrix} ae+bg & af+bh \\ ce+dg & cf+dh \end{pmatrix}.$$

This might be a good time to point out that often we are quite casual about making distinctions in either notation or language between operations on sets that are in fact not at all the same operation. For instance, in Example 6 we use the same symbol, a plus sign (+), to denote both the operation of matrix addition and the operation of addition of real numbers; moreover, we normally refer to each of these operations simply as "addition" as we do in this example and in Example 3.

With these six examples of rings well in hand, we are now ready for the formal definition of a ring. Our definition will lay down the list of axioms that any set with two operations must satisfy in order to attain the status of being called a ring. As you read this list of axioms, you might want to pause in turn and think about what each axiom says in the context of each of our six examples.

Definition 1.1. *A **ring** is a set R together with two operations on R (addition and multiplication) such that:*

1. addition is associative — that is, for all $a, b, c \in R$

$$a + (b + c) = (a + b) + c;$$

2. *addition is commutative — that is, for all a, b ∈ R*

$$a + b = b + a;$$

3. *R has a zero element — that is, there is an element 0 in R such that, for all a ∈ R*

$$a + 0 = a;$$

4. *for every a ∈ R, there is an element −a in R such that*

$$a + (-a) = 0;$$

5. *multiplication is associative — that is, for all a, b, c ∈ R*

$$a(bc) = (ab)c;$$

6. *multiplication is distributive over addition — that is, for all a, b, c ∈ R*

$$(a + b)c = ac + bc \quad and \quad a(b + c) = ab + ac.$$

The previous six axioms define a ring, *but we will want to concern ourselves in this book only with rings that satisfy two additional axioms. Thus, a* **commutative ring with identity** *is a ring R such that:*

7. *multiplication is commutative — that is, for all a, b ∈ R*

$$ab = ba;$$

8. *R has a multiplicative identity — that is, there is an element 1 in R such that for all a ∈ R*

$$a \cdot 1 = a.$$

It will be extremely important to remember that throughout the rest of this book the word *ring* will always mean *commutative ring with an identity element*. This should cause no confusion, but should always be kept firmly in mind, since the theory of noncommutative rings has quite a different character from commutative ring theory. Note that we have already seen one example of a noncommutative ring, the ring of *2 by 2 matrices*, $\mathbf{M}_2$, defined in Example 6, and that the set of even integers, $2\mathbf{Z}$, forms a commutative ring without an identity element.

Verifying that a given set together with two operations of addition and multiplication is in fact a ring — that is, that it satisfies all eight axioms — can be a long and tedious process (see Problem 1.7). Note also that the zero element mentioned in Axiom (3) can take different forms depending on the ring in question. For the rings $\mathbf{Z}$, $\mathbf{Q}$, and $\mathbf{R}$ the zero element is just the number 0. For the ring $\mathbf{C}$ it is the complex number $0 + 0i$ which we also usually denote more simply by 0. Similarly, for the ring of polynomials, $\mathbf{Z}[x]$, the zero element is the constant polynomial 0. For $\mathbf{M}_2$ it is the zero matrix $\begin{pmatrix} 0 & 0 \\ 0 & 0 \end{pmatrix}$. You should make sure that, for each of these examples of rings, you can identify the additive inverse mentioned in Axiom (4), as well as the multiplicative identity of Axiom (8).

Given that the eight ring axioms stated above were motivated by our desire to have rings behave algebraically just as we expect, it is not at all surprising that the following three familiar rules of algebra still hold in an arbitrary ring R:

1. $0a = 0$, for all $a \in R$;
2. $(-a)b = -(ab)$, for all $a, b \in R$;
3. $a(b - c) = ab - ac$, for all $a, b, c \in R$.

The verification of these rules from the axioms is left to you (see Problem 1.2).

The Definition of a Subring

It is frequently the case that we wish to focus our attention on a particular subset of a ring. For example, within the ring $\mathbf{R}$ of all real numbers we may wish to deal with the integers. The point here is that the integers themselves form a ring, and this ring shares with the larger ring $\mathbf{R}$ the operations of addition and multiplication, as well as having the same identity. In such a situation, we say that the subset in question is a *subring* of the larger ring.

As mentioned above, it is would be tedious always to have to check in complete detail that a given subset of a ring is itself a ring in order to know it is a subring. Fortunately, in the context we are discussing, there is a shortcut which we will adopt as our definition of a subring. Then I will leave to you in Problem 1.13 the one-time-only details of showing that this shortcut is in fact equivalent to the notion of a subring given in the preceding paragraph.

Using this shortcut, then, in order to verify that a given subset of a ring is indeed a subring, all that needs to be done is to check that the subset contains the identity of the larger ring, that the subset is *closed* under addition and multiplication — that is, if you add or multiply two elements of the subset then you get an element of the subset — and, finally, that the subset contains additive inverses — that is, for each element in the subset, its additive inverse is also in the subset.

Definition 1.2. *A subset S of a ring R is a **subring** of R if S is closed under the addition and multiplication operations of R, contains additive inverses, and contains the (multiplicative) identity of R.*

Problems

1.1   Show that in a ring the zero element, the multiplicative identity, and
        additive inverses are each unique — that is, there is only one element
        that behaves like 0, only one element that behaves like 1, and for each
        element a only one element that behaves like $-a$.

1.2   Use the eight ring axioms to prove the three familiar rules of algebra
        (1)–(3) listed on page 7. (Of course, $b - c$ is simply a convenient
        shorthand for $b + (-c)$.)

1.3   Let R be a ring, and let $a, b \in R$. Prove that $(-a)(-b) = ab$.

1.4   In this problem we see that the ordinary rules for exponents we are
        familiar with still work perfectly in a ring. Let R be a ring and let $a \in R$.
        We can inductively define powers of a as follows:

$$a^0 = 1 \quad \text{and} \quad a^n = a^{n-1}a \quad \text{for } n > 0.$$

        Use induction on n (fixing m as necessary) to prove that:

(i) $(ab)^n = a^n b^n$;   (ii) $a^m a^n = a^{m+n}$;   (iii) $(a^m)^n = a^{mn}$,

        for any non-negative integers m and n.
            Note that we have not defined a^{-1} since in an arbitrary ring a given
        element a may or may not have a multiplicative inverse. For example,
        the element 2 does not have a multiplicative inverse in the ring $\mathbf{Z}$.
        Therefore, the symbol a^{-1} should only be written when you are sure
        that the element a does in fact have a multiplicative inverse in the ring.

1.5   Let R be a ring such that $1 = 0$, where 0 is the zero element and 1 is the
        multiplicative identity in the ring. Show that R consists of just a single
        element. (By the way, you should convince yourself that, conversely,
        the *set* $R = \{0\}$ where the operations of addition and multiplication are
        defined by $0 + 0 = 0$ and $0 \cdot 0 = 0$ is actually a ring by verifying all the
        axioms.) This rather trivial, yet not altogether uninteresting, example
        of a ring is called the *zero ring*.

1.6   We define the set $\mathbf{Z} \times \mathbf{Z}$ to be the set of all ordered pairs of integers —
        that is, $\mathbf{Z} \times \mathbf{Z} = \{ (a, b) \mid a, b \in \mathbf{Z} \}$. Show how to make $\mathbf{Z} \times \mathbf{Z}$ into a ring
        by suitably defining the operations of addition and multiplication.
        What is the zero element of this ring? What is the multiplicative
        identity?

1.7   You may want to skip this exercise. It is long and tedious, but you should probably do it anyway, just so that you know you can go through the details of verifying the ring axioms when necessary. Assuming that the set of real numbers **R** is a ring, show that the complex numbers form a ring with operations as defined in Example 4.

1.8   We quite naturally denote the set of even integers by 2**Z**. Is 2**Z** a subring of **Z**?

1.9   Do the rationals form a subring of the reals?

1.10   Does the set of all numbers of the form $a + b\sqrt{3}$ where a and b are rational numbers form a subring of the reals?

1.11   Do the reals form a subring of the complex numbers?

1.12   Does the set of all numbers of the form $a + bi$ where a and b are integers form a subring of the complex numbers?

1.13   Let S be a subset of a ring R. Show that if S is a subring by the definition on page 7, then S is itself a ring.

2

Ideals and Quotient Rings

Ideals

As we continue our study of commutative ring theory we shall see that one of the concepts that makes the notion of a ring so useful is that of an *ideal*. This concept was introduced by Richard Dedekind, a student of Gauss. The name Dedekind is perhaps familiar to you, since he was the first to give a successful interpretation of irrational numbers, using what we now call Dedekind cuts.

Definition 2.1. *An **ideal** of a ring R is a subset I of R that is itself a group under addition and is such that if $r \in R$ and $a \in I$, then $ra \in I$.*

Let's break this definition apart into manageable pieces. To say that I is a group under addition means that I contains 0, the zero element of the ring, that I is *closed* under addition (that is, if $a, b \in I$ then $a + b \in I$), that I has additive inverses (that is, if $a \in I$ then $-a \in I$), and that addition in I is associative (which is certainly the case since addition is already associative in the ring R). If a subset of a ring is to be an ideal, then, it must first satisfy these four properties concerning the additive structure of the ring R. But, an *ideal* must also satisfy an extra — extremely strong — condition with respect to multiplication. I like to think of this property of an ideal as the *absorptive* property: not only must I be closed with respect to multiplication within I, but it must actually *absorb* multiplication from the ring R in the sense that if $a \in I$, then whenever we multiply a by any element $r \in R$ the product ra stays in I.

It is useful to contrast the notion of an ideal with that of a subring. First, an ideal must be closed under multiplication, so it is almost a subring, except that it need not contain 1, the identity element. On the other hand, a subring is almost an ideal, since it is a subgroup and is closed under multiplication, except that it need not have the stronger property of absorbing multiplication from the ring. Two examples will help to clarify this: the even integers are an ideal of **Z**, but not a subring, while the integers are a subring of the rationals, but not an ideal. You should pause here to make sure you believe these claims (see Problems 2.1 and 2.2).

Fortunately, in order to show that a given non-empty subset I is an ideal of a ring R, there is again a shortcut we can use; in fact, to show that I is an ideal we need only show that

1. if $a, b \in I$, then $a - b \in I$; and
2. if $r \in R$ and $a \in I$, then $ra \in I$.

Condition (1) ensures that I is a group and condition (2) says that I absorbs multiplication from R. In Problem 2.9 you are asked to verify that conditions (1) and (2) are in fact sufficient conditions for a set to be an ideal.

Let us look at some examples.

Example 1
The set $2\mathbf{Z}$ of even integers is an ideal of the ring $\mathbf{Z}$ of integers. In order to verify this it is enough to note that the difference of any two even integers is again an even integer, so condition (1) holds, and to note that if r is *any* integer and a is an even integer, then ra is also an even integer, so condition (2) holds as well.

Example 2
Let I be an ideal of a ring R. Suppose $1 \in I$: what can we say about the ideal I? It turns out that I must be the entire ring R. If we take an arbitrary element $r \in R$, then, since $1 \in I$, we have that $r = r \cdot 1 \in I$ by condition (2), which means that I in fact contains all the elements in R, thus $I = R$. This will be a very important idea for us: *whenever an ideal contains the identity element it must be the entire ring.*

We have already encountered the polynomial ring $\mathbf{Z}[x]$ in Example 5 in Chapter 1, consisting of all polynomials whose coefficients are integers. Obviously, in this example, there was nothing special about using the integers as coefficients other than the fact that we were able to add and multiply these integers whenever needed as part of the polynomial operations of addition and multiplication. In other words, the coefficients could have come from any ring, and they would work just as well in a polynomial ring. This leads us to the following more general definition of a polynomial ring.

Let R be a ring. The *polynomial ring over R* — that is, the ring of polynomials with coefficients from R — is denoted by $R[x]$. A typical element f of the polynomial ring $R[x]$ is, of course, a polynomial and looks like

$$a_0 + a_1 x + a_2 x^2 + \cdots + a_n x^n,$$

where $a_i \in R$ for each i. The term a_0 in the polynomial f is called the *constant term* and, provided that $a_n \neq 0$, n is the *degree* of the polynomial.

Example 3

Let R be a ring and let I be the set of all polynomials in the polynomial ring $R[x]$ with zero constant term — that is, if $f \in I$ then $f = a_1x + a_2x^2 + \cdots + a_nx^n$. Then, the set I is an ideal of $R[x]$. This is easy to verify. It is clear that the difference of two polynomials with zero constant term is again a polynomial with zero constant term, so condition (1) is satisfied. Now if $f \in R[x]$ and $g \in I$, then we can write

$$f = a_0 + a_1x + \cdots + a_nx^n \quad \text{and} \quad g = b_1x + \cdots + b_mx^m,$$

and it is clear that the polynomial fg has zero constant term, since

$$fg = a_0b_1x + \cdots + a_nb_mx^{n+m}.$$

Thus, $fg \in I$ and condition (2) is satisfied, and I is an ideal of $R[x]$.

Example 4

Let R be a ring and let J be the set of all polynomials in $R[x]$ having only *even* powers of x — that is, all polynomials of the form

$$a_0 + a_2x^2 + a_4x^4 + \cdots + a_{2n}x^{2n}.$$

Then J is a subring of $R[x]$, but not an ideal. J is not an ideal because it does not absorb multiplication by arbitrary elements of the ring $R[x]$: for example, if you multiply the above polynomial by the polynomial x you get a polynomial outside of J. (Alternatively, we could argue that since J contains the constant polynomial 1, which after all is the identity element of $R[x]$, we have seen that J must be the entire ring $R[x]$, which clearly it is not.) However, J is a subring because it contains the identity 1, is closed under addition and multiplication, and has additive inverses.

We now come to perhaps our most important example of an ideal:

Example 5

Let R be a ring and let $a \in R$. Let I be the set of all *multiples* of a — that is,

$$I = \{ra \mid r \in R\}.$$

Then I is an ideal of R. We must first show that the difference of any two elements of I is also in I. So we take two multiples $ra, sa \in I$, where $r, s \in R$, and observe that $ra - sa = (r - s)a \in I$, since $r - s \in R$, so their difference is also a multiple of a. Thus, I is a subgroup of R. We must now verify the absorptive property. Let $s \in R$ and let $ra \in I$, where $r \in R$. Then $s(ra) = (sr)a \in I$, since $sr \in R$. So, I absorbs multiplication from R (in this case, multiplication by the element s). Thus I is an ideal of R.

In fact, this example is important enough to merit its own definition:

Definition 2.2. *Let R be a ring and let $a \in R$. Then the ideal $I = \{ra \mid r \in R\}$ is called a* **principal ideal**. *We say that the ideal I is* **generated by a** *and usually write the ideal I as (a), or occasionally as Ra or aR if we wish to emphasize that we are multiplying the element a by elements of a particular ring R. (The notation $\langle a \rangle$ is also sometimes used for this ideal.)*

Example 6
The ideal I of Example 3 is in fact simply the principal ideal (x) generated by the polynomial x. This is probably clear to you, but I will prove it anyway in order to illustrate a useful technique for showing that two sets are equal. The technique consists of showing in turn that each of the two sets I and (x) are contained in one another. Hence, they are equal. First, we let $f \in I$ — that is, f is a polynomial with zero constant term. We can then factor an x out of f and write $f = g \cdot x$ for some $g \in R[x]$. (If $f = a_1 x + \cdots + a_n x^n$, then $g = a_1 + a_2 x + \cdots + a_n x^{n-1}$.) Thus, we have shown that $I \subseteq (x)$. Now we prove the reverse inclusion. Let $f \in (x)$; then $f = g \cdot x$ for some $g \in R[x]$. (In this case, $g = a_0 + \cdots + a_n x^n$, so $f = a_0 x + \cdots + a_n x^{n+1}$.) Clearly, then, f has zero constant term, and so $f \in I$. Thus we have shown that $(x) \subseteq I$. We conclude that $I = (x)$, as claimed.

Dedekind — in addition to giving the first successful description of irrational numbers — identified conditions (1) and (2) on page 12 as being of fundamental importance. Ideals can be thought of as a generalization of the concept of number. In particular, we can identify a number with the set of other numbers that it divides — that is, with the *ideal generated by that number*. This important idea motivates our next example:

Example 7
In this example we see that the *only* ideals of the integers $\mathbf{Z}$ are in fact principal ideals — that is, ideals such as (6), or (-2), or (1729). The argument goes like this. Let I be an ideal of $\mathbf{Z}$. One possiblity is that $I = (0)$, the ideal consisting of just the single element 0 and appropriately called the *zero ideal*. Otherwise, the ideal I must contain at least one positive integer (since if $n \in I$ for some nonzero integer n,

then $-n \in I$ as well, and one of the integers n or $-n$ is positive). Let n be the *least* positive integer contained in I. We claim that $I = (n)$. Since $n \in I$, any multiple of n is also an element of I, so $(n) \subseteq I$. To verify the reverse inclusion we let $a \in I$. By the division algorithm, we can write $a = qn + r$, where $q, r \in \mathbf{Z}$ and $0 \leq r < n$. (In other words, r is the remainder we get when we divide a by n.) But, $r = a - qn \in I$, since $a \in I$ and $qn \in I$. Since n was the *least* positive integer in I, we conclude that $r = 0$, and so $a = qn$, which means that $a \in (n)$. Hence, $I \subseteq (n)$. It follows that $I = (n)$, as claimed. Therefore, *all* ideals of $\mathbf{Z}$ are principal.

One word of caution about notation before we turn our attention to quotient rings. Our standard convenient notation for a principal ideal is (a), which unfortunately can sometimes be ambiguous. For example, if we simply write an ideal generated by 2 as (2), this would mean the ideal $2\mathbf{Z}$ if we are working in the integers, but an entirely different ideal if we are working in the rationals $\mathbf{Q}$. This is the reason for the more precise notation Ra or aR for principal ideals. (Of course, in this book, where we are dealing only with commutative rings, there is no need to use both Ra and aR for principal ideals. They are both mentioned here merely to serve as a gentle reminder that in the world of noncommutative rings things can be different.) Thus, in the two cases above we could write respectively $2\mathbf{Z}$ and $2\mathbf{Q}$ in order to remove any ambiguity.

Quotient Rings

In this section we turn our attention to the way in which ideals can be used to create additional examples of rings. If you have studied group theory you know that the *normal subgroups* play a particularly important role in forming what are called *quotient groups*. In ring theory ideals play this same vital role.

The basic idea is to begin with a given ring and an ideal of that ring, and then to use that ideal to create an entirely new ring from the given ring. Now, the elements of this new ring will actually turn out to be subsets of the original ring. The only real trick is to figure out how to "add" and "multiply" these subsets. It is no accident that ideals give rise precisely to the subsets that can be manipulated algebraically in the way we want. These special subsets — which will be called *cosets* — are described in the following definition:

Definition 2.3. *Let I be an ideal of a ring R. A* **coset** *of I is a set of the form*

$$a + I = \{a + i \mid i \in I\},$$

where $a \in R$. The element a is called a **representative** *of the coset $a + I$.*

Example 8

Let $I = (5)$ be an ideal of $\mathbf{Z}$, the ideal consisting of all multiples of 5. Then we can list the following cosets of the ideal I:

$$0 + (5) = \{\ldots, -10, -5, 0, 5, 10, \ldots\},$$

$$1 + (5) = \{\ldots, -9, -4, 1, 6, 11, \ldots\},$$

$$2 + (5) = \{\ldots, -8, -3, 2, 7, 12, \ldots\},$$

$$3 + (5) = \{\ldots, -7, -2, 3, 8, 13, \ldots\},$$

$$4 + (5) = \{\ldots, -6, -1, 4, 9, 14, \ldots\}.$$

Moreover, these are the *only* cosets of the ideal I, since using any other number as a representative results in one of the five cosets listed above — for example, $8 + (5)$ is the same coset as $3 + (5)$, and $1729 + (5)$ is the same coset as $4 + (5)$. Note that a given coset can be represented by any of its elements, and only by these elements. Note also that these five cosets are disjoint, and that each element of $\mathbf{Z}$ is in one of these five cosets. In such a situation we say that these cosets form a *partition* of $\mathbf{Z}$.

The cosets of an ideal I will themselves form the elements of a ring, called the *quotient ring*. We already have the elements in place for such a ring, namely the cosets. So, in Example 8 there will be exactly five elements in the quotient ring. What is missing is the algebraic structure; in other words, how do we add and multiply cosets? In Example 8 it is pretty obvious how we might decide to add, say, the two cosets $2 + (5)$ and $4 + (5)$ — we get $6 + (5)$, which is just the coset $1 + (5)$; to multiply them we get $8 + (5)$, which is just the coset $3 + (5)$. But, before we can define these operations in a general setting, we need to learn more about cosets.

The Basic Properties of Cosets

Let I be an ideal of a ring R. Let $a, b \in R$. Then the cosets of I have the following basic properties:

1. $a + I = I$ if and only if $a \in I$;
2. $a + I = b + I$ if and only if $a - b \in I$;
3. $a + I = b + I$ if and only if $(a - b) + I = I$;
4. the two cosets $a + I$ and $b + I$ are either disjoint or equal.

Proofs

(1). First, we assume that $a + I = I$, and show that $a \in I$. But $a = a + 0 \in a + I$, so $a \in I$. Conversely, we assume that $a \in I$, and show that $a + I = I$. In order to show that these two sets are equal, we show that each is contained in the other. First, we take $a + b \in a + I$, where $b \in I$ — that is, we take a typical element of $a + I$. We must show that $a + b \in I$. But I is an ideal, and both a and b are elements of I, so $a + b \in I$. Thus, $a + I \subseteq I$. On the other hand, let $c \in I$. We must show that $c \in a + I$. Since I is an ideal, $-a \in I$, and so $c - a \in I$. Hence, we have $c = a + (c - a) \in a + I$, as desired. Thus, $I \subseteq a + I$. We conclude that $a + I = I$.

(2). First, we assume that $a + I = b + I$. Since

$$a = a + 0 \in a + I = b + I,$$

we see that $a \in b + I$. Therefore, we can write $a = b + i$ for some $i \in I$. But then $a - b = i \in I$, and $a - b \in I$, as desired. Conversely, assume that $a - b \in I$. Then $a - b = i$, for some $i \in I$. In order to show that the two sets $a + I$ and $b + I$ are equal, we show that each is contained in the other. First, we take $a + c \in a + I$, where $c \in I$. So

$$a + c = (b + i) + c = b + (i + c) \in b + I,$$

since $i + c \in I$. Thus, $a + I \subseteq b + I$. At this stage we could prove containment in the other direction with a similar argument. However, it is easier to notice that $a - b \in I$ implies that $b - a \in I$. Hence, the preceding argument goes through with b and a interchanged. Thus, $b + I \subseteq a + I$. We conclude that $a + I = b + I$.

(3). This follows immediately from (1) and (2).

(4). Suppose that $a + I$ and $b + I$ are not disjoint. Let x be an element in their intersection. Then we can write $x = a + i$, for some $i \in I$, and also write $x = b + j$, for some $j \in I$. Thus, $a + i = b + j$. Hence, $a - b = j - i \in I$, since $j, i \in I$. By property (2), $a - b \in I$ implies that $a + I = b + I$ and the two cosets are equal. We conclude that if these two cosets are not disjoint, then they are equal. This completes the proofs of the four basic properties of cosets.

We are now ready to define the key operations of *coset addition* and *coset multiplication*. Keep in mind that the elements we are trying to add and multiply are actually cosets.

Definition 2.4. *Let I be an ideal of a ring R. The operation of* **addition** *on the set of cosets of I is given by*

$$(a + I) + (b + I) = (a + b) + I,$$

for $a, b \in R$. The operation of **multiplication** *on the set of cosets of I is given by*

$$(a + I)(b + I) = ab + I,$$

for $a, b \in R$.

Note that these definitions appear to depend on the particular representatives a and b used for the cosets. It therefore needs to be verified that these operations are in fact *well defined*, but rather than delay the development of the main idea of quotient rings any further, I leave this important detail for you to investigate, if you are interested (see Problem 2.23).

It should come as no great surprise to you by now that — in this very natural way — an ideal I of a ring R gives rise to a *ring of cosets* in which the operations of addition and multiplication are defined as above. This simple and beautiful idea is one of the most fruitful in commutative ring theory.

Definition 2.5. *Let I be an ideal of a ring R. The* **quotient ring** *of R by I is the ring of cosets of I, where addition and multiplication of cosets are defined as above. We denote this ring by R/I, which we read as "R mod I."*

The easy — but necessary — details of showing that the quotient ring R/I as defined above is actually a ring are left to you (see Problem 2.24). We emphasize once again that the elements of this quotient ring are the cosets $a + I$, where $a \in R$. The zero element of the quotient ring is $0 + I$ (that is, the ideal I itself), and the multiplicative identity is $1 + I$.

Example 8 revisited
Let $I = (5)$ be the ideal generated by 5 in the ring of integers **Z**. The quotient ring $\mathbf{Z}/I$ consists of five elements — namely, $0 + (5)$, $1 + (5), 2 + (5), 3 + (5)$, and $4 + (5)$. We can illustrate the operations of addition and multiplication in this ring:

$$(0 + I) + (2 + I) = 2 + I,$$

$$(2 + I) + (4 + I) = 6 + I = 1 + I,$$

$$(1 + I)(3 + I) = 3 + I,$$

$$(2 + I)(4 + I) = 8 + I = 3 + I.$$

Obviously, all we are really doing here is arithmetic modulo 5. The ring $\mathbf{Z}/(5)$ is often called the *ring of integers modulo 5*, and is sometimes written as $\mathbf{Z}_5$.

The term "quotient ring" and the notation "R/I" can be somewhat misleading since they have nothing to do with quotients or fractions of numbers. Rather, both the notation and the term are meant to suggest that the ideal I *factors* or *partitions* or *divides* the ring R into cosets, each having the same size as I. However, as long as you read R/I as "R mod I", and remember that the elements of this ring are cosets, there is little danger of confusing the concept of a quotient ring with the still-to-come notion of a *ring of fractions*.

Problems

2.1   We saw in Example 1 that the set $2\mathbf{Z}$ of even integers is an ideal of the ring $\mathbf{Z}$ of integers. Show that $2\mathbf{Z}$ is not a subring of $\mathbf{Z}$.

2.2   The ring of integers $\mathbf{Z}$ is a subring of the rationals $\mathbf{Q}$. Show that $\mathbf{Z}$ is not an ideal of $\mathbf{Q}$.

2.3   Find a subring of $\mathbf{Z}[x]$ that is not an ideal of $\mathbf{Z}[x]$.

2.4   Find an ideal of $\mathbf{Z}[x]$ that is not a subring of $\mathbf{Z}[x]$.

2.5   Find all of the ideals of the ring $\mathbf{Z}/(12)$.

2.6   Find all of the ideals of the ring $\mathbf{Z} \times \mathbf{Z}$.

2.7   Let I and J be ideals of a ring R. Prove that the intersection $I \cap J$ is also an ideal of R.

2.8   We saw in Example 7 that *all* ideals of $\mathbf{Z}$ are principal. Illustrate this by finding an integer n such that $(8) \cap (14) = (n)$.

2.9   Prove that a nonempty subset I of a ring R is an ideal if and only if the set I satisfies conditions (1) and (2) given on page 12.

2.10   Let R be a ring. Prove that the intersection of *any* collection of ideals of R is also an ideal of R — that is, if $\{I_\lambda\}_{\lambda \in \Lambda}$ is a collection of ideals of R, then

$$\bigcap_{\lambda \in \Lambda} I_\lambda$$

        is an ideal of R.

2.11   Give a specific example to show that the union $I \cup J$ of two ideals of a ring R need not be an ideal of R. (Hint: try the ring $R = \mathbf{Z}$.)

2.12   Let I and J be ideals of a ring R. Obviously, if either of these ideals is contained in the other, then their union is also an ideal since their union is just the larger of the two ideals. Prove that if, on the other hand, neither I nor J is contained in the other, then $I \cup J$ is not an ideal.

2.13   Let R be a ring and let $a \in R$. Prove that (a) — that is, the principal ideal generated by a — is contained in any ideal I which contains the element a.

2.14   Let R be a ring and let $a \in R$. Prove that (a) — that is, the principal ideal generated by a — is the intersection of *all* ideals containing the element a. That is, prove that

$$(a) = \bigcap_{a \in J} J.$$

Because of this fact, we often think of the principal ideal (a) as the *smallest* ideal containing the element a.

2.15   We can define the ideal (a, b) that is *generated by two elements a and b* of a ring R as follows:

$$(a, b) = \{ ra + sb \mid r, s \in R \}.$$

Show that (a, b) is an ideal.

2.16   Let R be a ring and let $a, b \in R$. Prove that (a, b) — the ideal generated by a and b — is the intersection of *all* ideals containing both of the elements a and b. In other words, prove that

$$(a, b) = \bigcap_{a, b \in J} J.$$

Because of this fact, we often think of the ideal (a, b) as the *smallest* ideal containing both of the elements a and b.

2.17   We saw in Example 7 that *all* ideals of **Z** are principal. Illustrate this by finding an integer n such that $(8, 14) = (n)$, where $(8, 14)$ is the ideal in **Z** generated by 8 and 14 (see Problem 2.15).

2.18   Show that the ideal (x, y) generated by the polynomial x and the polynomial y is not a principal ideal of $\mathbf{Z}[x, y]$, the ring of polynomials in two variables with integer coefficients.

2.19   Find an ideal of $\mathbf{Z}[x]$ generated by two elements that is not a principal ideal.

2.20   A **field** is a nonzero ring (meaning, as always, a commutative ring with identity) in which each nonzero element a has a multiplicative inverse a^{-1}. In other words, in a field you can always divide by any element except 0. For example, **Q** is a field, but **Z** is not.
        Let R be a nonzero ring. Prove that
        R is a field if and only if the only ideals of R are (0) and R itself.

2.21   Consider the set $C(\mathbf{R})$ of all *continuous* real-valued functions on the real
        line. For any two functions $f, g \in C(\mathbf{R})$ we can define functions $f + g$
        and fg as follows:

$$(f + g)(x) = f(x) + g(x) \quad \text{and} \quad (fg)(x) = f(x)g(x).$$

We say that two functions $f + g$ and fg are *defined pointwise*. Prove that
with these two operations of addition and multiplication $C(\mathbf{R})$ is a ring.

2.22   Let $I = \{\, f \in C(\mathbf{R}) \mid f(17) = 0 \,\}$ — that is, I is the set of all continuous
        functions whose value at 17 is 0. Prove that I is an ideal of $C(\mathbf{R})$.

2.23   The definition of *coset addition* is given on page 17. What this
        definition says is: in order to add two cosets we take a representative a
        of one of the cosets, and a representative b of the other coset; then
        their sum $a + b$ represents the coset which is the *sum* of the two
        original cosets. But what if we should happen to take different
        representatives at the start? Would we still get the same coset for an
        answer? In particular, if a' is a different representative of the first coset,
        and b' is a different representative of the second coset, must $a + b$ and
        $a' + b'$ represent the same coset? If the answer is ever no, then *our
        definition of coset addition makes no sense whatsoever*. If the answer is
        always yes, then we can say that that addition of cosets is **well
        defined**. Therefore, in order to show that addition of cosets is
        well-defined, we must show that whenever

$$a + I = a' + I \quad \text{and} \quad b + I = b' + I,$$

then

$$(a + b) + I = (a' + b') + I.$$

Prove that addition and multiplication of cosets are both well-defined
operations.

2.24   Let I be an ideal of a ring R. Prove that the quotient ring R/I is in fact a
        ring.

2.25   Let $\mathbf{Z}$ be the ring of integers, and let n be a positive integer. Recall from
        Problem 2.20 that a field is a nonzero ring in which each nonzero
        element has a multiplicative inverse. Prove that

$$\mathbf{Z}/(n) \text{ is a field if and only if } n \text{ is a prime number.}$$

3

Prime Ideals and Maximal Ideals

Divisibility

Ideals are a generalization of the concept of number; it is therefore not surprising that those ideals that correspond to prime numbers are of particular importance. But how are we to capture the notion of *primeness* in an arbitrary ring? It turns out that the key to this question is divisibility.

Definition 3.1. *Let* $m, n \in \mathbf{Z}$. *We say that* m **divides** n *if* $n = rm$ *for some* $r \in \mathbf{Z}$, *and in this case we write* $m|n$, *a notation which is read as "m divides n."*

In the language of ideals, if m divides n it means precisely that $n \in (m)$, from which it follows that $(n) \subseteq (m)$. This argument is reversible, so we are able to summarize the way in which the divisibility is mirrored in the ideal structure of the ring $\mathbf{Z}$ as follows:

$$m|n \text{ if and only if } (n) \subseteq (m).$$

In other words, a statement about the divisibility of integers has been translated into a statement about containment of the corresponding ideals. Note in particular that as long as the numbers in question are positive, the smaller number actually corresponds to the larger ideal.

Returning to our task of capturing the notion of primeness in an arbitrary ring, we further observe that if p is a prime number, then p has the following well-known divisibility property, where a and b are integers:

$$\text{if } p|ab, \text{ then } p|a \text{ or } p|b.$$

That is, if a prime number divides a product of two integers, then it must divide one of the two numbers. This is not true for composite numbers — for example, 15 divides $6 \cdot 10$, but 15 does not divide either 6 or 10. This fundamentally important divisibility property of prime numbers is equivalent to the original notion of primeness, and also turns out to be the one that is most suitable for generalization.

Definition 3.2. *Let P be a proper ideal of a ring R — that is, P ≠ R. Then the ideal P is* **prime** *if for a, b ∈ R*

$$ab \in P \text{ implies } a \in P \text{ or } b \in P.$$

In other words, a proper ideal is prime if, whenever the product of two elements is in the ideal, then one of the two elements must be in the ideal. For example, in **Z** the ideal (15) is not prime since $6 \cdot 10 \in (15)$, but neither 6 nor 10 is in the ideal (15), whereas (7) is a prime ideal of **Z** since if $ab \in (7)$ then $7|ab$, and so $7|a$ or $7|b$, and $a \in (7)$ or $b \in (7)$.

Since our definition of prime ideals was motivated entirely by the notion of primeness in the integers, it should come as no surprise to you that the nonzero prime ideals in **Z** are precisely those principal ideals that are generated by a prime number.

Example 1
Let *n* be a positive integer. Then

(*n*) is a prime ideal of **Z** if and only if *n* is a prime number.

In order to verify this, we first assume that (*n*) is a prime ideal. We use contradiction to prove that *n* is a prime number. Assume that *n* is not prime, and let $n = ab$ be a nontrivial factorization of *n* — that is, where $a, b \in \mathbf{Z}$ such that $1 < a, b < n$. Then $ab \in (n)$, so $a \in (n)$ or $b \in (n)$, since (*n*) is a prime ideal. But this is impossible, since $1 < a, b < n$. Therefore, *n* is a prime number.

Conversely, assume that *n* is a prime number. Let $ab \in (n)$, with $a, b \in \mathbf{Z}$. We must show that $a \in (n)$ or $b \in (n)$. Since $ab \in (n)$, we can write $ab = rn$ for some $r \in \mathbf{Z}$. Thus, $n|ab$, but *n* is a prime number, so $n|a$ or $n|b$. That is, $a = sn$ or $b = tn$, for some $s, t \in \mathbf{Z}$. Therefore, $a \in (n)$ or $b \in (n)$, and we conclude that (*n*) is a prime ideal, as desired.

Integral Domains

One of the convenient properties of the ring **R** of real numbers — and one which we certainly tend to take for granted — is the fact that the product of two nonzero real numbers is never equal to zero. This simple but exceedingly important fact is the basis for the standard technique of solving equations by factoring. For example, in order to solve the equation

$$x^2 + x - 6 = 0,$$

we can factor to get

$$(x - 2)(x + 3) = 0.$$

Since this product is 0, one of the two factors must be 0, so we write

$$x - 2 = 0 \quad \text{or} \quad x + 3 = 0,$$

which yields the solution

$$x = 2 \quad \text{or} \quad x = -3.$$

The preceding method relies heavily on the fact that if the product of two factors is 0, then one of the two factors is 0. However, this is not always the case in an arbitrary ring. Consider, for example, the ring of integers modulo 6 — that is, the ring $\mathbf{Z}/(6)$. (Recall that there are six elements in this ring: the cosets $0 + (6)$, $1 + (6)$, $2 + (6)$, $3 + (6)$, $4 + (6)$, and $5 + (6)$.) Then, for example, the product of the two elements $3 + (6)$ and $4 + (6)$ is given by

$$(3 + (6))(4 + (6)) = 12 + (6) = 0 + (6),$$

but $0 + (6)$ is the zero element of this ring. In other words, the product is 0, but neither of the factors is 0. This admittedly awkward situation gives rise to the following definition.

Definition 3.3. *Let R be a ring and let $a \in R$. Then the element a is said to be a* **zero-divisor** *if there is a nonzero element $b \in R$ such that $ab = 0$.*

For example, we saw above that $3 + (6)$ is a zero-divisor in the ring $\mathbf{Z}/(6)$. (Incidentally, we saw at the same time that $4 + (6)$ is a zero-divisor as well.) Note that the zero element 0 is a zero-divisor in any nonzero ring. On the other hand, the ring of integers $\mathbf{Z}$ and the ring of real numbers $\mathbf{R}$ have no zero-divisors except 0. It was precisely this property of $\mathbf{R}$ that we used in our example illustrating the technique of solving equations by the method of factoring. Rings such as $\mathbf{Z}$ and $\mathbf{R}$, which have no zero-divisors except 0, are worthy of a special name.

Definition 3.4. *A nonzero ring R is called an* **integral domain** *or, often more simply, a* **domain** *if R has no zero-divisors other than 0.*

Example 2
Recall from Problem 2.20 in Chapter 2 that a *field* is a nonzero ring in which each nonzero element a has a multiplicative inverse a^{-1}. We

claim that

any field is an integral domain.

Let K be a field. Suppose that $a \in K$ is a zero-divisor in K; then $ab = 0$ for some $b \neq 0 \in K$. However, K is a field, so b has an inverse b^{-1} in K, and we can write

$$a = a \cdot 1 = a(bb^{-1}) = (ab)b^{-1} = 0 \cdot b^{-1} = 0.$$

So the only zero-divisor in K is 0; that is, K is an integral domain.

Example 3
Let K be a field. Then $K[x]$, the ring of polynomials with coefficients from K, is an integral domain. We give a proof by contradiction. Suppose that $f \neq 0$ is a zero-divisor in $K[x]$. Let g be a nonzero polynomial in $K[x]$ such that $fg = 0$. We write

$$f = a_0 + a_1 x + a_2 x^2 + \cdots + a_n x^n \text{ and } g = b_0 + b_1 x + b_2 x^2 + \cdots + b_m x^m,$$

where $a_i, b_i \in K$, for each i, and $a_n, b_m \neq 0$. Since $fg = 0$, we have

$$0 = fg = a_0 b_0 + (a_0 b_1 + a_1 b_0)x + \cdots + a_n b_m x^{n+m}.$$

Therefore, each of the coefficients of this product must be 0. In particular, then, $a_n b_m = 0$. But neither a_n nor b_m equals 0, which contradicts the fact that the field K is an integral domain. We conclude that $K[x]$ has no zero-divisors other than 0, and so $K[x]$ is an integral domain, as claimed.

We saw above that the ring $\mathbf{Z}/(6)$ is not an integral domain. Obviously, this is because 6 is a composite number. On the other hand, we know that if p is a prime number, then $\mathbf{Z}/(p)$ is a field (see Problem 2.25), and hence $\mathbf{Z}/(p)$ is an integral domain. This fundamental relationship between prime ideals and their quotient rings — namely, that prime ideals give rise to quotient rings that are always integral domains — is a relationship that holds in general, and is of such importance that we make it our next theorem.

Theorem 3.1. *Let P be an ideal of a ring R. Then*

P is a prime ideal if and only if R/P is an integral domain.

Proof

First, assume that P is a prime ideal. We must show that R/P is an integral domain — that is, that R/P has no zero-divisors other than the zero element. Recall that the zero element of the ring R/P is $0 + P$, which we usually write as just plain P. Suppose that $a + P$ is a zero-divisor in R/P, where $a \in R$. Then there is some nonzero element $b + P$, where $b \in R$ (note that since $b + P$ is nonzero in R/P, that means $b + P \neq P$, that is, $b \notin P$) such that $(a + P)(b + P) = 0 + P = P$. Thus, $ab + P = P$, and so $ab \in P$. Now, P is a prime ideal, so $a \in P$ or $b \in P$. But, we know that $b \notin P$, so $a \in P$. Therefore, $a + P = P$. We have thus shown that the only zero-divisor in R/P is P, and hence that R/P is an integral domain.

Conversely, assume that R/P is an integral domain. We must show that P is a prime ideal. Let $a, b \in R$ such that $ab \in P$. Then we can write $(a + P)(b + P) = ab + P = P$. Therefore, since R/P is an integral domain with zero element P, we conclude that either $a + P = P$ or $b + P = P$. Hence, $a \in P$ or $b \in P$, and P is a prime ideal. This completes the proof of the theorem.

This theorem is a nice illustration of a strategy that is very useful in commutative ring theory: since various properties of a ring R are often reflected in corresponding properties of quotient rings R/I — and vice versa — we can play these properties off against one another. In other words, in order to learn something about a ring R, it may be helpful to change the location of the discussion to a ring R/I instead.

Suppose, for example, that we are curious about the nature of a particular ideal I in a ring R. If we can somehow decide that R/I is an integral domain, then Theorem 3.1 tells us that I is a prime ideal. On the other hand, what if we are able to learn instead that R/I is in fact a field? What does this tell us about the ideal I? The answer to this question is the topic of the next section.

Maximal Ideals

There is one particular type of prime ideal that is especially important. These are the *maximal* ideals. They are called maximal because they are not contained in any other proper ideal — that is, they are as large as they can be and still be a proper ideal. (Note that this concept is different from saying that an ideal has a *maximum* number of elements. In fact, it is possible for a ring to have maximal ideals with different numbers of elements, that is, of different sizes.)

Definition 3.5. *Let M be a proper ideal in a ring R — that is, $M \neq R$. Then, M is* **maximal** *if the only ideal properly containing M is the whole ring R itself.*

Thus, M is a maximal ideal of R if there are no other ideals contained between M and R. In other words, if N is an ideal such that

$$M \subseteq N \subseteq R,$$

then

$$N = M \quad \text{or} \quad N = R.$$

As we see in the next example, this will be our basic strategy for showing that an ideal M is maximal — that is, we will take an ideal N between M and R and show that $N = M$ or $N = R$; or, alternatively, we will take an ideal N *properly* between M and R and reach a contradiction.

Example 4

Let us show that the ideal $M = (7)$ is a maximal ideal of **Z**. Suppose that N is an ideal of **Z** *properly* contained between (7) and **Z** — that is, $(7) \subset N \subset \mathbf{Z}$. Then there exists an integer $n \in N \setminus (7)$. Since $n \notin (7)$, 7 and n are relatively prime, which means that the greatest common divisor of 7 and n is 1, there are integers x and y such that $1 = 7x + ny$. (This is an easy consequence of the well-known Euclidean algorithm for finding the greatest common divisor of two integers.) But, $7 \in N$ and $n \in N$, so $1 \in N$, and it follows that $N = \mathbf{Z}$. We conclude that (7) is a maximal ideal.

Example 5

On the other hand, the ideal (6) is obviously not a maximal ideal of **Z**, since $(6) \subset (2) \subset \mathbf{Z}$.

As you may have guessed by now,

all maximal ideals are prime ideals.

This can be proved directly from the definitions, and you will be asked to do this in Problem 3.9. We will shortly be able to offer a much more elegant proof using quotient rings. But first we need to introduce a bit of notation.

We have already encountered the notion of an ideal that is generated by elements of ring R: for example, a principal ideal (a) generated by a single element, or even the ideal (a, b) generated by two elements. We have also seen that there are two ways of viewing each of these ideals.

Thus, the ideal (a) can be thought of as the collection of all multiples ra of the element a, or it can be thought of as the intersection of all ideals containing the element a (see Problem 2.14). Similarly, the ideal (a, b) can be thought of as all linear combinations $ra + sb$ of the elements a and b, or it can be thought of as the intersection of all ideals containing the two elements a and b (see Problems 2.15 and 2.16). Obviously, this procedure can be generalized to any number of generators, which leads to the following definition.

Definition 3.6. *Let S be a subset of a ring R. Then the **ideal generated by S** is denoted by (S) and is the intersection of all the ideals that contain S — that is,*

$$(S) = \bigcap_{J \supseteq S} J.$$

That this intersection is in fact an ideal follows from Problem 2.10. Alternatively, we can also give an elementwise description of the ideal generated by S:

$$(S) = \{r_1 s_1 + r_2 s_2 + \cdots + r_n s_n \mid r_i \in R, s_i \in S\}.$$

In other words, (S) is just the set of all linear combinations of elements from the set S. I leave to you in Problem 3.7 the necessary step of showing that this useful alternate description coincides with the above definition.

Note that we write (a) instead of $(\{a\})$, and (a, b) instead of $(\{a, b\})$. Similarly we will write (I, a) instead of $(I \cup \{a\})$ for the ideal generated by I and a. For instance, the proof in Example 4 that (7) is a maximal ideal of **Z** amounted to showing that if we *add* any element n to the ideal (7), then the resulting ideal $((7), n)$ is the *entire* ring **Z**.

Since it will often be useful to have a description of the elements of an ideal of the form (I, a), it will also be left for you to show, in Problem 3.8, that

$$(I, a) = \{i + ra \mid i \in I \text{ and } r \in R\}.$$

With this notation in hand, we are now ready to prove the theorem which does for maximal ideals what Theorem 3.1 did for prime ideals:

Theorem 3.2. *Let M be an ideal of a ring R. Then,*

M is a maximal ideal if and only if R/M is a field.

Proof

First, assume that M is a maximal ideal. Then $M \neq R$, so R/M is a nonzero ring. We show that R/M is a field by demonstrating as we must that any nonzero element of R/M has an inverse. Let $a + M$ be a nonzero element of R/M — that is, $a \in R \setminus M$. Let $N = (M, a)$. Since $a \notin M$, the ideal N *properly* contains M, but M is maximal, so we conclude that $N = R$. Thus, $1 \in N = (M, a)$, and we can write $1 = m + ra$, for some $m \in M$ and some $r \in R$. We claim that $r + M$ is the desired inverse of $a + M$, which follows immediately from

$$(r + M)(a + M) = ra + M = 1 - m + M = 1 + M.$$

Therefore, R/M is a field.

Conversely, suppose that R/M is a field. We must show that M is a maximal ideal. Since R/M is nonzero, $M \neq R$. Suppose that N is an ideal that properly contains M — that is, $N \supset M$. We must show that N is the entire ring R. Let $a \in N \setminus M$. Then $a + M$ is a nonzero element of the field R/M and, as such, has an inverse $b + M$, where $b \in R$. Thus,

$$1 + M = (b + M)(a + M) = ba + M.$$

Therefore, by Property (2) of cosets on page 16, $1 - ba \in M$, and we can write $1 - ba = m$, for some $m \in M$. Since $a, m \in N$, we finally have $1 = ba + m \in N$, and we can conclude that $N = R$, showing that M is a maximal ideal. This completes the proof of the theorem.

Since, as we saw in Example 2, any field is an integral domain, we can now use the strategy of passing to quotient rings to prove without any effort whatsoever the following important result:

Corollary 3.1. *Any maximal ideal is also a prime ideal.*

Proof

Let M be a maximal ideal of a ring R. Then, by Theorem 3.2, R/M is a field. But, any field is an integral domain, so R/M is an integral domain, and so, by Theorem 3.1, M is a prime ideal. This completes the proof.

Not all prime ideals are maximal, however, as the following examples show.

Example 6

The zero ideal (0) of $\mathbf{Z}$ is a particularly simple example of a prime ideal that is not maximal. To see that (0) is prime, let $ab \in (0)$, which means that $ab = 0$, and so $a = 0$ or $b = 0$, and $a \in (0)$ or $b \in (0)$, thus showing (0) to be prime. To see that (0) is not maximal, we simply observe that $(0) \subset (2) \subset \mathbf{Z}$.

Example 7

A less trivial — but equally easy — example of a nonmaximal prime ideal is given by the ideal (x) in the polynomial ring $\mathbf{Z}[x, y]$. To see that (x) is prime, let $fg \in (x)$; then $x|fg$, so $x|f$ or $x|g$, and so $f \in (x)$ or $g \in (x)$ and (x) is prime. To see that (x) is not maximal, we simply observe that $(x) \subset (x, y) \subset \mathbf{Z}[x, y]$.

Given the fundamental importance of prime and maximal ideals in commutative ring theory, it would be nice to know whether an arbitrary ring must necessarily contain either a maximal or a prime ideal. In the next chapter we shall see — although at considerable effort — that every nonzero ring contains at least one maximal ideal, and hence at least one prime ideal.

Problems

3.1   Let P be a prime ideal of a ring R — that is, P is a proper ideal such
        that if $ab \in P$ for $a, b \in R$, then $a \in P$ or $b \in P$. Prove that if $abc \in P$
        for $a, b, c \in R$, then at least one of the three elements must be
        in P. More generally, use induction to prove that if $a_1 a_2 a_3 \cdots a_n \in P$
        for $a_i \in R$, then at least one of the factors a_i must be in P.

3.2   Let R be a ring. Prove directly — that is, without using Theorem 3.1 —
        that

        (0) is a prime ideal of R if and only if R is an integral domain.

3.3   A subset of a ring is said to be a **multiplicative system** if it is closed
        under multiplication and contains the multiplicative identity 1. Let
        R be a ring and P an ideal of R. Prove that

        P is a prime ideal if and only if $R \setminus P$ is a multiplicative system.

3.4   Find all prime ideals and all maximal ideals in $\mathbf{Z}/(12)$.

3.5   Find all prime ideals and all maximal ideals in $\mathbf{Z} \times \mathbf{Z}$.

3.6   Let I and J be two ideals of a ring R, neither contained in the other.
        Prove that the ideal $I \cap J$ is *not* a prime ideal.

3.7   Let S be a subset of a ring R. Show that the set

        $$\{r_1 s_1 + \cdots + r_n s_n \mid r_i \in R, s_i \in S\}$$

        is actually an ideal. Then argue that this set coincides with our
        definition of (S), the ideal generated by S.

3.8   Let I be an ideal of a ring R, and let $a \in R$. Prove that the set

        $$\{i + ra \mid i \in I \text{ and } r \in R\}$$

        is an ideal of R, and hence is the ideal (I, a) generated by I and a.

3.9   Prove directly from the definitions — that is, without using Theorems
        3.1 and 3.2 — that any maximal ideal is prime. (Hint: let M be a
        maximal ideal; suppose that $ab \in M$, but $a \notin M$; show that $b \in M$.)

3.10   In this problem we see that for rings without a multiplicative identity a maximal ideal need not necessarily be prime. This is one of the reasons that in this book we require our rings to have a multiplicative identity.
        Consider the *group of integers modulo 7* — that is, the integers $0, 1, 2, 3, 4, 5, 6$, where *addition* is defined mod 7; for example, $3 + 6 = 2$. We can make this group into a "ring" R by by defining multiplication as follows:

$$ab = 0, \text{ for all } a, b \in R.$$

This "ring" has no identity, but it satisfies all our other ring axioms. Show that R has a maximal ideal that is not prime.

3.11   We saw in Example 2 that any field is an integral domain. In this problem we prove a partial converse. Let D be a *finite* integral domain — that is, D has a finite number of elements. Prove that D is a field.

3.12   We saw in Examples 6 and 7 that a prime ideal need not be maximal. In this problem we see that in some rings this converse to the corollary of Theorem 3.2 is true. Let R be a *finite* ring — that is, a ring with a finite number of elements. Prove that any prime ideal of R is also a maximal ideal.

3.13   The **product** of two ideals I and J is defined to be the ideal generated by all products ab, where $a \in I$ and $b \in J$. That is, if $S = \{ab \mid a \in I, b \in J\}$, then the product of I and J is given by

$$IJ = (S).$$

We can also view this product as all sums of such products — that is,

$$IJ = \{a_1 b_1 + a_2 b_2 + \cdots + a_n b_n \mid a_i \in I \text{ and } b_i \in J\}.$$

Our definition of a prime ideal is given in terms of the elements of the ring. In this problem we give a purely ideal-theoretic version of what it means for an ideal to be prime. Let P be an ideal in a ring R. Prove that P is prime if and only if

$$IJ \subseteq P \text{ implies } I \subseteq P \text{ or } J \subseteq P$$

for any ideals I and J of R.

3.14   Let D be a ring. Prove that

   D is an integral domain if and only if $D[x]$ is an integral domain.

   In particular, this result shows that if K is a field, then $K[x]$ is an integral domain, as we saw in Example 3.

3.15   Let D be a ring. Prove that D is an integral domain if and only if $D[x, y]$ is an integral domain.

   More generally, prove that D is an integral domain if and only if $D[x_1, x_2, \ldots, x_n]$ is an integral domain. Conclude that if K is a field, then $K[x_1, x_2, \ldots, x_n]$ is an integral domain.

4

Zorn's Lemma and Maximal Ideals

Partial Orders

Our primary goal in this chapter is to show that every ring has a maximal ideal. A proof of this very simple and apparently obvious fact will require the use of something fundamental known as Zorn's lemma. Since it happens to be the case that Zorn's lemma is one of the indispensable tools for the modern mathematician, we will take this opportunity to make a rather roundabout detour from our study of commutative rings in order to explore not only the way Zorn's lemma is used in commutative ring theory but also its larger place in mathematics. We will have a few other occasions to use Zorn's lemma later in the book as well; still, an impatient reader may reasonably choose to bypass this detour and pick up the main route again in the next chapter.

It will be Zorn's lemma that allows us to assert with complete confidence that a particular ring that we might have in our hands really does contain a maximal ideal. The situation is not unlike that when Hilbert proved the existence of a finite basis for invariants. We need not actually construct a maximal ideal in our ring — though certainly we often do this as well — we are frequently merely satisfied just to know that such an ideal exists.

Since our goal in this chapter is the existence of *maximal* ideals, let us first turn to a general discussion of maximal elements. Loosely speaking, something is said to be *maximal* if there is nothing bigger. So, for example, Mount Everest is maximal. Of course, one problem is that you don't always have maximal elements in a given system. For example, there is no maximal integer — that is, there is not an integer n with the property that no integer m is larger than n. As we have seen, Dedekind's idea was to look at the integers in a new way and to focus on ideals rather than numbers as the primary objects. Thus, if we use *set inclusion* to measure the size of ideals within the integers, then we do have maximal elements, after all. For example, the ideal (3) will be said to be bigger than the ideal (6) because $(6) \subset (3)$. In fact, in this sense, (3) is maximal, as we have seen, since no other proper ideal is bigger. On the other hand, some ideals, such as (3) and (5), can't be compared with one another by means of set inclusion, so we can't say that either one is larger than the other.

In order to state and subsequently use Zorn's lemma in our search for maximal elements, we need to make precise exactly how we compare things. This leads us to the notion of a partial order:

Definition 4.1. *A **partial order** on a set S is a relation ≤ on S which is*

1. *reflexive — that is, for all $x \in S$,*

$$x \leq x;$$

2. *antisymmetric — that is, for all $x, y \in S$,*

$$x \leq y, y \leq x \text{ implies } x = y;$$

3. *transitive — that is, for all $x, y, z \in S$,*

$$x \leq y, y \leq z \text{ implies } x \leq z.$$

Let us look at several examples of partial orders. In each case the order is a very natural one, and so it is quite easy to verify the three axioms for a partial order. I leave this to you. Whenever a partial order on a given set seems to be the most natural one, we call it the *standard partial order*.

Example 1
Let S be the set of integers **Z**. We define the standard partial order ≤ on S as follows, where $x, y \in S$:

$$x \leq y \text{ if and only if } y - x \text{ is non-negative.}$$

Example 2
Let S be a collection of sets. We define the standard partial order ≤ on S as follows, where $A, B \in S$:

$$A \leq B \text{ if and only if } A \subseteq B.$$

Example 3
Let S be a collection of sets. This time we will define another quite natural order ≤ on S, called *reverse inclusion*, as follows, where $A, B \in S$:

$$A \leq B \text{ if and only if } A \supseteq B.$$

Example 4

Let S be the set $C(\mathbf{R})$ of continuous real-valued functions on the real line. We define the standard partial order $\leq$ on S as follows, where $f, g \in S$:

$$f \leq g \text{ if and only if } f(x) \leq g(x) \text{ for all } x \in \mathbf{R}.$$

In other words, $f \leq g$ if the graph of g never dips below the graph of f.

It needs to be emphasized that in each case we are using the familiar symbol $\leq$ to represent a partial order in a purely formal way, and you should not attribute any meaning to the symbol $\leq$ other than what is given to it by the definition of the particular partial order. In Example 1 the meaning given to $\leq$ is of course the meaning you are used to. In Example 2 it admittedly feels a bit strange to write

$$\{0, 2, 4\} \leq \{0, 1, 2, 3, 4, 5\},$$

but it is still quite reasonable. In Example 3, however, it seems almost perverse to write

$$\{0, 1, 2, 3, 4, 5\} \leq \{0, 2, 4\},$$

but that is what $\leq$ means for this particular partial order.

Note also that in the first partial order above, any two elements of the set S can be compared — that is, for any two integers x and y, either $x \leq y$ or $y \leq x$. This is not the case for the other partial orders. For example, given two sets A and B, it is certainly possible for neither to be contained in the other, so that we have neither $A \leq B$ nor $B \leq A$. Similarly, the functions f and g given by $f(x) = x^2$ and $g(x) = x^3$ are not comparable, since their graphs cross. This, of course, is why we use the term *partial* order. In the particularly nice situation where any two elements in a set *can* be compared to one another we have the following:

Definition 4.2. *A partial order $\leq$ on a set S is a* **total order**, *or a* **linear order**, *if for any $x, y \in S$, either $x \leq y$ or $y \leq x$. In particular, if $\leq$ is a partial order on a set S and C is a subset of S, then we say that the set C is a* **chain** *if $\leq$ is a total order on C.*

Since Zorn's lemma involves chains, let us look at a few examples:

Example 5

The standard order on $\mathbf{R}$, the set of real numbers, is a total order. In particular, this means that any subset of $\mathbf{R}$ is a chain.

This is in fact how you should think of an arbitrary chain, with the elements of the chain all strung out before you in a row, like the real numbers.

Example 6

Let S be a set with the standard partial order given by inclusion — that is, the order given in Example 2. Let $A_1, A_2, A_3, \ldots$ be subsets of S. The set

$$C = \{A_1, A_1 \cup A_2, A_1 \cup A_2 \cup A_3, \ldots\}$$

is a chain, since given any two sets in C, one is contained in the other.

We can make the total order of such a chain stand out more clearly by writing the chain as follows:

$$A_1 \subseteq A_1 \cup A_2 \subseteq A_1 \cup A_2 \cup A_3 \subseteq \cdots.$$

This is an example of an *ascending* chain.

Example 7

Let $C([0, 1])$ be the set of all continuous real-valued functions on the closed interval $[0, 1]$ with the standard partial order given in Example 4. Then the set of functions

$$\cdots \leq x^3 \leq x^2 \leq x$$

is a chain, which we have displayed in the usual way with bigger things on the right (don't forget: these functions are only defined on $[0, 1]$).

However, we will often write such a chain as

$$x \geq x^2 \geq x^3 \geq \cdots,$$

since it is a *descending* chain.

These last two examples — as well as the term *chain* itself — give the impression that chains are countable, but they need not be. For example, the irrationals $\mathbf{R} \setminus \mathbf{Q}$ form an uncountable chain in $\mathbf{R}$. In fact, we need to be very careful not to assume countability in a proof using chains.

We are getting closer to Zorn's lemma, but we still need a few more concepts.

Definition 4.3. *Let $\leq$ be a partial order on a set S. Let A be a subset of S. An* **upper bound** *of the set A is an element $s \in S$ such that $a \leq s$, for all $a \in A$.*

An upper bound of a set A is simply an element that is bigger than (or possibly equal to) every element in the set A. A set A might have many upper bounds, or a single upper bound, or even possibly no upper bounds at all. For example, the set of prime numbers has no upper bound in the integers since there is no largest prime, whereas the set of negative integers has infinitely many upper bounds in the integers, since any positive integer is an upper bound (so are 0 and -1, for that matter).

Also, an upper bound of a set A may or may not itself be in the set for which it is an upper bound. For example -1 is an upper bound for the set of negative integers and it is a negative integer, but 2 is also an upper bound for the set of negative integers and it is not a negative integer.

Moreover, whether or not a set A has an upper bound depends on the set S. For example, let

$$A = \{1/2, 2/3, 3/4, 4/5, \ldots\}.$$

If A is considered as a subset of S, where S is the *closed* interval $[0, 1]$, then A has an upper bound in S, namely 1, but if A is considered as a subset of T where T is the *open* interval $(0, 1)$, then A has no upper bound in T.

Zorn's lemma concerns the existence of maximal elements in a partially ordered set, so we need a precise definition of a maximal element, a definition which, at this point, you could probably provide on your own.

Definition 4.4. *Let $\leq$ be a partial order on a set S. An element $m \in S$ is a* **maximal element** *of S if there is no element $s \in S$, $s \neq m$, such that $m \leq s$.*

In other words, a maximal element of a set S is simply an element of S such that no other element in S is larger. In particular, though, a maximal element of a set must actually be in the set. Some partially ordered sets have maximal elements and some don't. For example, with the standard order for real numbers, the closed interval $S = [0, 1]$ has a maximal element, whereas the open interval $T = (0, 1)$ does not. Some partially ordered sets even have more than one maximal element! For example, the set S of proper ideals in the ring $\mathbf{Z}$ has lots of maximal elements — namely, the maximal ideals $(2), (3), (5), \ldots$ corresponding to the prime numbers. This last example also explains our normal use of the term *maximal ideal* since what we mean when we say an ideal is a maximal ideal is precisely that it is a maximal element in the set S of proper ideals with the standard partial order of set inclusion.

Recall that our goal is to show the existence of maximal ideals. Zorn's lemma provides us with a way of guaranteeing the existence of maximal elements in a partial order. It turns out that all we have to do is check that any chain has an upper bound:

Theorem 4.1 (Zorn's lemma). *Let $\leq$ be a partial order on a non-empty set S. If every chain in S has an upper bound in S, then S contains a maximal element.*

Now the truth of Zorn's lemma is not at all obvious. It is hardly even plausible. You certainly have every reason to be expecting a proof at this point. But, instead, I will briefly describe the state of affairs regarding the use of Zorn's lemma in mathematics, after which you will perhaps understand why a proof has been omitted. Before we do that, however, let's look at the long-promised application of Zorn's lemma to the question of the existence of maximal ideals in rings. This is not only a very important result in commutative ring theory, but also an absolutely typical illustration of the use of the indispensable mathematical tool which is Zorn's lemma.

Theorem 4.2. *Every nonzero ring has at least one maximal ideal.*

Proof
Let R be a nonzero ring. Let S be the set of all proper ideals of R, together with the standard partial order of set inclusion. The set S is non-empty since (0) is a proper ideal of R — this is an innocent looking but easily forgotten step that is often crucial in any application of Zorn's lemma. Clearly, a maximal element of the set S is a maximal ideal of R (which, of course, is why we chose the set S in the way we did).

In order to apply Zorn's lemma we need only show that any chain in S has an upper bound in S. Let C be a chain is S. Since S is a set of ideals, C is in fact a chain consisting of ideals, and we have a rather obvious candidate for an upper bound for C: the union U of *all* the ideals in the chain C. Clearly, any ideal in the chain C is contained in U, so U is the upper bound we are looking for, as long as U is in the set S — that is, provided that U is in fact a proper ideal of R.

So, we need to show that U is a proper ideal of R. First, let $a, b \in U$. We must show that $a - b \in U$. Since $a \in U$ and U is the union of all the ideals in C, the element a is in some ideal I in the chain C. Similarly, $b \in J$ for some ideal J in C. But, by definition, the chain C is totally ordered, so I and J are comparable. Either $I \subseteq J$ or $J \subseteq I$. We may as well assume that $I \subseteq J$. Therefore, $a \in I \subseteq J$, but $b \in J$, so $a - b \in J$, since J is an ideal.

Thus $a - b \in U$, since U is the union of all the ideals in the chain C and, as such, contains J.

Next, let $a \in U$, and let $r \in R$. We must show that $ra \in U$. But $a \in I$ for some ideal I in the chain C. Thus $ra \in I$, since I is an ideal, and so $ra \in U$. Therefore, U is an ideal of R.

Finally, U is a *proper* ideal of R since $1 \notin U$. This is because 1 is not in any of the ideals in the chain C, since all of these ideals are proper ideals.

Thus, the chain C has an upper bound in S, and so, by Zorn's lemma, S has a maximal element, which in turn is a maximal ideal of the ring R. This completes the proof.

It is worth noting that the upper bound U that we found in the above proof is merely an upper bound for the chain C and need not itself be a maximal element. We still do not have a way to construct maximal ideals; we just know they exist.

The key to any successful use of Zorn's lemma is making appropriate choices of the set S and the partial order $\leq$. These need to be chosen so that when Zorn's lemma gives you a maximal element within this framework, this maximal element is in fact what you were looking for. For example, in the above proof, a maximal element of S actually turns out to be a maximal ideal, which is exactly what we were looking for.

You may be bothered that I have shown no inclination whatsoever to prove Zorn's lemma itself, in spite of having just made use of it to prove Theorem 4.2. If so, you should be somewhat comforted to learn that Zorn's lemma is actually equivalent to the following statement, a statement which when you think about it does appear to be obvious!

Axiom of Choice. *Let S_λ be a non-empty set for each λ in an index set Λ. Then there is a function f defined on the set Λ such that $f(\lambda) \in S_\lambda$ for each $\lambda \in \Lambda$.*

This axiom, like any good axiom, simply states something that seems perfectly obvious and thus requires no proof — namely, that you can *choose* an element from each of the sets S_λ. This seems absolutely clear. Therefore, since Zorn's lemma can be proved using the Axiom of Choice, we should also believe Zorn's lemma.

If it indeed seems obvious to you that we can always choose an element from each set S_λ — and I for one agree that it does — consider for a moment the following illustration due to Bertrand Russell.

Take an uncountably infinite set of pairs of shoes and an uncountably infinite set of pairs of socks. Our goal is to choose one shoe from each pair of shoes, and also to choose one sock from each pair of socks. This sounds easy enough, and it is, for the shoes! We could, for example, simply choose all the left shoes. In this way we get a set of shoes, one shoe from each pair. (This amounts to defining the function f in the Axiom of Choice by $f(\lambda) = $ the left shoe in the pair of shoes S_λ.)

But now try to choose a sock from each pair. We need the Axiom of Choice to do so! That is, we cannot give a concrete description of our choice function (as we could for shoes). All we can do is assume that such a choice could be made (and again, I for one am willing to assume this).

I hope the reader is beginning to feel the ever-so-slight rumbling of the philosophical ground upon which we mathematicians stand when using tools such as Zorn's lemma or the Axiom of Choice in order to demonstrate the existence of some mathematical object we might need at a given moment, say, a maximal ideal in a commutative ring. So, a little caution is called for.

Apparently, the first person to realize that something like the Axiom of Choice was needed was Giuseppe Peano in 1890. He was seeking an existence proof in the study of differential equations, and wrote: "However, since one cannot apply infinitely many times an arbitrary law by which one assigns to a class an individual of that class, we have formed here a definite law by which, under suitable assumptions, one assigns to every class of a certain system an individual of that class."

The Axiom of Choice was fully formulated by Bertrand Russell in 1906, but its place in set theory was not to be resolved for over half a century.

By the end of the nineteenth century, mathematicians felt that they had at long last shored up the foundations of mathematics. As Henri Poincaré put it to the Second International Congress of Mathematicians in 1900: "Mathematics ... has been arithmetized ... we may say today that absolute rigor has been obtained."

But within two years of Poincaré's announcement, new cracks in the foundations began to appear as a result of the now famous paradox of Bertrand Russell. Russell's paradox has many forms (see Problems 4.2 and 4.3), the most popular of which was given by Russell in 1919. A barber in a village shaves every man in the village who does not shave himself. Who shaves the barber? If the barber does not shave himself, then he *does* shave himself. On the other hand, if he does shave himself, then he does *not* shave himself. It became clear that set theory, as devised by Georg Cantor and gradually accepted by mathematicians, needed revising.

The axiomatic system that is used by most mathematicians these days is called Zermelo-Fraenkel set theory. Thus, there are two central questions for us at this point. Is the Axiom of Choice consistent with Zermelo-Fraenkel set theory? That is to say, can we include it as an axiom without fear of contradiction? In 1938, Kurt Gödel proved the Axiom of Choice to be consistent. (Incidentally, he proved this while also proving the more difficult fact that the *continuum hypothesis* is consistent. The continuum hypothesis claims that there is no subset of the real numbers, the continuum, whose cardinality is strictly between that of the integers and that of the reals; one of Cantor's great achievements, of course, was to prove that the cardinality of the reals is strictly greater than the cardinality of the integers.)

The other central question for us is whether the Axiom of Choice is independent of the axioms of Zermelo-Fraenkel set theory, or might the Axiom of Choice actually be a theorem within Zermelo-Fraenkel set theory? This was not settled until 1963, when Paul Cohen proved that the Axiom of Choice is independent. Thus, the standard footing for a mathematician today is Zermelo-Fraenkel set theory together with the Axiom of Choice. However, if one is to be scrupulously honest, it is also customary to call attention to any specific use that is made of the Axiom of Choice.

There are many logically equivalent forms of the Axiom of Choice. Zorn's lemma is but one of these. We conclude our lengthy detour by mentioning in passing one other form of the Axiom of Choice that is frequently encountered. A set is said to be **well-ordered** if we can give it a total order such that every non-empty subset has a least element. The Axiom of Choice — and hence, Zorn's lemma — is also equivalent to the following:

Well-Ordering Principle. *Any set can be well-ordered.*

Problems

4.1   Use Zorn's lemma to prove that every proper ideal of a ring is contained in a maximal ideal. (Hint: let I be an ideal of a ring R; the key is to choose carefully a set S of ideals such that when Zorn's lemma gives you a maximal element of S, this will turn out to be a maximal ideal containing I.)

4.2   An adjective is called *autological* if it can be applied to itself. For example, "short" is a short word, so the adjective "short" is autological. An adjective that is not autological is called *heterological*. For example, "blue" is heterological since the word "blue" is not blue. Is the adjective "heterological" autological or heterological?

4.3   There are two kinds of sets, those that are members of themselves, and those that are not. For example, the set $\{1, 2, 3\}$ is not a member of itself, since the only members are 1, 2, and 3. On the other hand, the set of all sets is a member of itself, since it is a set. Here is another form of Russell's paradox. Let S be the set of all sets that are not members of themselves. Is the set S a member of itself? Or not?

4.4   Let $\{P_\lambda\}_{\lambda \in \Lambda}$ be a set of prime ideals in a ring R that form a chain under inclusion. Show that the intersection of this chain of prime ideals is a prime ideal.

4.5   Let I be an ideal of a ring R that is contained in a prime ideal Q. Use Zorn's lemma to prove that there is a prime ideal P of R such that

$$I \subseteq P \subseteq Q,$$

and there are *no* prime ideals between I and P. We will call such a prime P a **minimal prime over I**.

4.6   Prove that every nonzero ring has at least one **minimal prime ideal** — that is to say, a prime ideal that contains no other prime ideal.

5

Units and Nilpotent Elements

Units

If an ideal I of a ring contains the multiplicative identity 1, then we have seen that I must be the entire ring. We have used this extremely simple idea several times already — for example, in proving that all rings have maximal ideals and also in proving that an ideal is maximal if and only if its quotient ring is a field. More generally, there are other elements in a ring that — like the identity 1 — have this same property that they can be in an ideal only if the ideal is the entire ring. These elements are called the *units*.

Definition 5.1. *Let R be a ring. An element $u \in R$ is called a **unit** if it has an inverse — that is, if there exists an element $v \in R$ such that $uv = 1$.*

In particular, the identity 1 is a unit in any ring. In the ring of integers $\mathbf{Z}$, the only units are 1 and -1. However, at the other extreme, in a field every nonzero element is a unit. Since the terms unit and identity are quite similar, it is easy to confuse them, so you need to be a bit careful to keep them straight.

Now, as we have suggested, an ideal can contain a unit only if the ideal is the entire ring. In order to see this, suppose that u is a unit in a ring R and that $u \in I$ for some ideal I of R. Since u is a unit, it has an inverse $u^{-1} \in R$. We conclude that $1 = u^{-1}u \in I$. Hence, $I = R$. That is, any ideal that contains a unit must be the entire ring. This can be summarized by saying that if u is a unit, then $(u) = R$. On the other hand, if an element $a \in R$ is *not* a unit, then $(a) \neq R$, for otherwise, if $(a) = R$, then $1 \in (a)$, and $1 = ra$ for some $r \in R$, which is contrary to the fact that a is a non-unit. We conclude that for an element $u \in R$

$$u \text{ is a unit if and only if } (u) = R.$$

We can also reverse our perspective and focus instead on the nonunits of a ring. The following theorem gives a very nice characterization of the set of non-units of a ring in terms of the maximal ideals.

Theorem 5.1. *Let R be a ring. The union of all the maximal ideals of R is the set of non-units of R:*

$$\bigcup_{M \text{ maximal}} M = \text{set of non-units.}$$

Proof

Let $u \in R$. We have just shown that if u is a unit, then u is not contained in any proper ideal. In particular, then, if u is a unit, $u \notin M$ for any maximal ideal M of R (since, if $u \in M$, then $1 = u^{-1}u \in M$ implies that $M = R$). Hence, no unit is contained in the union of the maximal ideals. This shows that the union of the maximal ideals is contained in the set of non-units.

On the other hand, if an element u is not contained in the union of the maximal ideals, then $u \notin M$ for any maximal ideal, and so the ideal (u) is not contained in any maximal ideal. Thus, (u) is not a proper ideal, since any proper ideal is contained in a maximal ideal by Zorn's lemma (see Problem 4.1). Therefore, $(u) = R$, and u is a unit. This shows that the set of non-units is contained in the union of the maximal ideals. Since each of these two sets is contained in the other, this completes the proof.

Nilpotent Elements and the Nilradical

The foregoing theorem describes the elements that make up the union of the maximal ideals of a ring — namely, the non-units. We now give a description of the elements that make up the intersection of the prime ideals of a ring. These elements are the *nilpotent* elements.

Definition 5.2. *Let R be a ring. An element $a \in R$ is said to be **nilpotent** if $a^n = 0$ for some positive integer n. The set of nilpotent elements in a ring R is called the **nilradical** of R. We denote the nilradical by $\mathcal{N}$.*

In the next theorem we will see that in any ring

the nilradical is the intersection of the prime ideals of the ring.

Since the intersection of any collection of ideals is an ideal (see Problem 2.10), it will follow that the nilradical is an ideal. It is nonetheless worth our effort to go ahead at this time and give a direct proof that the nilradical $\mathcal{N}$ of a ring R is an ideal. This not only will give us some practice working with nilpotent elements, but will also give us a chance to use the binomial theorem and point out that this important theorem still works in any commutative ring.

As usual, there are two things to verify in order to show that the nilradical $\mathcal{N}$ is an ideal. Let $a, b \in \mathcal{N}$. This means that $a^n = 0$ and $b^m = 0$ for some positive integers n and m. We must show that $a - b$ is nilpotent. We will do this by showing specifically that $(a - b)^{n+m-1} = 0$. This follows by using the binomial theorem to expand $(a - b)^{n+m-1}$ to get

$$a^{n+m-1} - \binom{n+m-1}{1}a^{n+m-2}b^1 + \cdots + (-1)^{m-1}\binom{n+m-1}{m-1}a^n b^{m-1}$$

$$+ (-1)^m\binom{n+m-1}{m}a^{n-1}b^m + \cdots + (-1)^{n+m-1}b^{n+m-1}.$$

Each term of this expression is zero, since either the exponent of a is at least n (as it is for the terms in the first row above), or the exponent of b is at least m (as it is for the terms in the second row). Hence, $(a - b)^{n+m-1} = 0$, and $a - b$ is nilpotent, as desired.

Next, we let $a \in \mathcal{N}$, and let $r \in R$. We must show that ra is nilpotent. Since $a \in \mathcal{N}$, we can write $a^n = 0$ for some positive integer n. But then $(ra)^n = r^n a^n = 0$, so ra is nilpotent. Therefore, the nilradical $\mathcal{N}$ is an ideal.

We now prove the following striking theorem, which characterizes the set of nilpotent elements in a ring in terms of the prime ideals of the ring. Surprisingly, the proof of this theorem requires Zorn's lemma.

Theorem 5.2. *Let $\mathcal{N}$ be the nilradical of a ring R. Then $\mathcal{N}$ is the intersection of all prime ideals in R — that is,*

$$\mathcal{N} = \bigcap_{P \ prime} P.$$

Proof

Let $a \in \mathcal{N}$. We must show that a is an element of the intersection of all the prime ideals — that is, we must show that $a \in P$ for each prime ideal P. Let P be a prime ideal of R. Then $a^n = 0 \in P$ for some positive integer n, since $a \in \mathcal{N}$. It follows that $a \in P$, since P is a prime ideal (see Problem 3.1). Thus, $a \in \bigcap P$, the intersection being taken over all prime ideals P. So, the nilradical is contained in the intersection of all the prime ideals of R.

On the other hand, if $a \in R$ and $a \notin \mathcal{N}$, then we must show that there is a prime ideal P such that $a \notin P$, and hence that a is not an element of the intersection of *all* prime ideals of R. In order to do this we need to use Zorn's lemma.

As is always the case when using Zorn's lemma, the key is a careful selection of the set S. In this case, we want a maximal element of S to be a prime ideal P such that $a \notin P$. With this in mind, we let S be the set of all ideals I of R such that $a^n \notin I$ for all positive integers n — that is, neither a nor any of its powers a^n lie in I. First, we check that S is non-empty. Since $a \notin \mathcal{N}$, $a^n \neq 0$ for any n, and so $a^n \notin (0)$ for any n. Thus, the zero ideal (0) is an element of the set S, and we can be sure that S is non-empty.

Now, we order the set S by inclusion — that is, for two ideals I and J, $I \leq J$ if and only if $I \subseteq J$. In order to apply Zorn's lemma we must show that any chain of ideals in S has an upper bound in S. Let C be a chain of ideals in S. Then the union U of all the ideals in the chain C is itself an ideal of R and, furthermore, is an ideal in S, for if any power of a lies in the union U, then that same power lies in one of the ideals in the chain C, contrary to the assumption that the ideals of C are elements of the set S. Thus, the chain C has an upper bound in S — namely, its union. Hence, by Zorn's lemma, S has a maximal element, which we shall call P. (We are not saying that P is a maximal ideal — merely, that P is a maximal element in the partially ordered set S.)

Since P is an element of S, no power of a lies in P. In particular, $a \notin P$. Therefore, we are done as soon as we know that P is a prime ideal. In order to show this, let $x, y \in R$ such that $x, y \notin P$. Then it is sufficient to show that $xy \notin P$ (see Problem 3.3). Since $x \notin P$, the ideal (P, x) — that is, the ideal generated by P and x (see Problem 3.8) — is strictly greater than the ideal P, in other words, $P \subset (P, x)$. But P is *maximal* in the partially ordered set S, so (P, x), being bigger than P, cannot be in the set S. Thus, by the definition of the set S, there is a positive integer m such that $a^m \in (P, x)$. Similarly, $a^n \in (P, y)$ for some positive integer n. It easily follows that $a^{m+n} \in (P, xy)$, the ideal generated by the ideal P and the element xy (see Problem 5.9). Hence, $(P, xy) \notin S$, and so (P, xy) is strictly greater than P (which *is* in S) — that is, $P \subset (P, xy)$. Thus, $xy \notin P$, and we conclude that P is prime. Finally, then, since $a \notin P$, a prime ideal, we can conclude that $a \notin \bigcap P$, the intersection of all prime ideals of R. This completes the proof.

Note once again that the fact that the nilradical $\mathcal{N}$ of a ring is actually an ideal follows immediately from Theorem 5.2, since the intersection of a collection of ideals is an ideal.

Problems

5.1   Find all the units in $\mathbf{Z}/(12)$. Then verify that the result of Theorem 5.1 holds for this ring (see Problem 3.4).

5.2   Find all the units in $\mathbf{Z} \times \mathbf{Z}$. Then verify that the result of Theorem 5.1 holds for this ring (see Problem 3.5).

5.3   Find all the nilpotent elements in $\mathbf{Z}/(12)$. Then verify that the result of Theorem 5.2 holds for this ring (see Problem 3.4).

5.4   Find all the nilpotent elements in $\mathbf{Z} \times \mathbf{Z}$. Then verify that the result of Theorem 5.2 holds for this ring (see Problem 3.5).

5.5   Let R be a ring. Prove that each element of R is either a unit or a nilpotent element if and only if the ring R has a unique prime ideal. Give an example of such a ring.

5.6   Let a be a nilpotent element of a ring R. Prove that $1 + a$ is a unit of R. (Hint: you want to show that the expression $\frac{1}{1+a}$ makes sense in the ring R; one approach is to represent this expression as an infinite series; another approach is to use long division.)

5.7   The ring of **Gaussian integers** is the following subring of the complex numbers:

$$\mathbf{Z}[i] = \{a + bi \mid a, b \in \mathbf{Z}\}.$$

In this problem we find the units of the ring $\mathbf{Z}[i]$. In order to do this we define an integer-valued function N on $\mathbf{Z}[i]$ by

$$N(a + bi) = a^2 + b^2.$$

1. Show that if $z_1, z_2 \in \mathbf{Z}[i]$, then $N(z_1 z_2) = N(z_1)N(z_2)$.
2. Let $z \in \mathbf{Z}[i]$. Show that z is a unit if and only if $N(z) = 1$.
3. Find all the units of $\mathbf{Z}[i]$.

5.8   Let I and J be two ideals of a ring R. We define the **sum** $I + J$ to be the ideal consisting of all sums $a + b$, where $a \in I$ and $b \in J$ — that is,

$$I + J = \{a + b \mid a \in I, b \in J\}.$$

Show that, with this definition, the sum $I + J$ is in fact an ideal of R. Then, in particular, show that, for $x \in R$,

$$I + (x) = (I, x).$$

5.9   Let I be an ideal of a ring R, and let $x, y \in R$. We defined the *product* of two ideals in Problem 3.13. Show that the product of the two ideals $I + (x)$ and $I + (y)$ is contained in the ideal $I + (xy)$ (which by Problem 5.8 is the same as the ideal (I, xy)) — that is, show that

$$\left(I + (x)\right)\left(I + (y)\right) \subseteq I + (xy) = (I, xy).$$

Give an example to illustrate that the product of $I + (x)$ and $I + (y)$ can be *properly* contained in the ideal $I + (xy)$.

6

Localization

God made the integers. All else is the work of man.
  — Leopold Kronecker

The Rational Numbers and Equivalence Relations

It is of course well known to us that the rational numbers can be constructed from the integers — that is, the rationals are just fractions, or ratios, of integers. In fact, it is because of the utter naturalness of this construction that the Pythagoreans of the sixth century B.C. believed that rational numbers were the *only* numbers — hence, their deep philosophical shock when they discovered that not all lengths could be represented by rational numbers. For example, the diagonal of a unit square has length $\sqrt{2}$, which, as they could show, is not a rational number. It was not until the nineteenth century that the arithmetic of the mysterious *irrational* numbers was made precise by Dedekind.

While it is true that rational numbers are essentially just fractions, that is not quite right, since, for example, we do not distinguish between 3/4 and 6/8. So, the rationals are indeed fractions of integers, but using something slightly subtle that is called an *equivalence relation* — that is, we agree to make certain fractions, such as 3/4 and 6/8, equivalent. This leads to the following definition.

Definition 6.1. *A relation $\cong$ on a set S is called an* **equivalence relation** *if it is*

    1. *reflexive — that is, for all $x \in S$,*

$$x \cong x;$$

    2. *symmetric — that is, for all $x, y \in S$,*

$$x \cong y \text{ implies } y \cong x;$$

    3. *transitive — that is, for all $x, y, z \in S$,*

$$x \cong y, \ y \cong z \text{ implies } x \cong z.$$

Example 1

Let S be the set of all triangles in the plane. We define — usually subconsciously — the standard equivalence relation on S by saying that two triangles are *equivalent* if they are *congruent* — that is, all three of their sides and all three of their angles are respectively equal. In other words, where the triangles are located in the plane, or even if they are turned over, is declared to be irrelevant. It is easy to see that congruence of triangles is an equivalence relation by checking that the relation of congruence is reflexive, symmetric and transitive.

Example 2

Let S again be the set of all triangles in the plane. We define another standard equivalence relation on S by saying that two triangles are *equivalent* if they are *similar* — that is, all three of their angles are respectively equal. In other words, we are now saying that neither location nor size matters. Again it is easy to see that similarity of triangles is an equivalence relation.

Example 3

Let I be an ideal of a ring R. We can define an equivalence relation on R by saying that two elements of R are equivalent if they are in the same coset of I — that is, we define

$$a \cong b \text{ if and only if } a - b \in I,$$

for $a, b \in R$.

It is worth going through the details of showing that this relation is reflexive, symmetric, and transitive. Since $a - a = 0 \in I$ for all $a \in R$, $\cong$ is reflexive. For all $a, b \in R$, if $a - b \in I$, then $b - a = -(a - b) \in I$, so $\cong$ is symmetric. Finally, for all $a, b, c \in R$, if $a - b \in I$ and $b - c \in I$, then $a - c = (a - b) + (b - c) \in I$, so $\cong$ is transitive. Thus, $\cong$ is an equivalence relation.

Note that, for $I = (n)$ in $\mathbf{Z}$,

$$a \cong b \text{ if and only if } a - b \in (n),$$

for $a, b \in \mathbf{Z}$. We normally write this equivalence relation as $a \equiv b$ (mod n), and we say that a and b are congruent mod n. More generally then, we write $a \equiv b$ (mod I) if $a - b \in I$, and we say that a and b are congruent mod I.

Example 4

Let S be the set of fractions a/b such that $a, b \in \mathbf{Z}$ and $b \neq 0$. The standard equivalence relation on this set of fractions is one that we use without even thinking about it. The equivalence relation on fractions is given by

$$\frac{a}{b} \cong \frac{c}{d} \text{ if and only if } ad = bc$$

for $a, b, c, d \in \mathbf{Z}$ and $b, d \neq 0$. (You might want to go through the details of showing that this is an equivalence relation, especially since the algebra is a little tricky for transitivity.) This explains precisely what we are doing when we think of $3/4$ and $6/8$ as being the same number, because $3 \cdot 8 = 4 \cdot 6$. In fact, any fraction that is equivalent to $3/4$, such as $132/176$, is also thought of as representing the same rational number. Note that we are thinking of a *single* rational number as being *represented* by various fractions such as $3/4$ or $6/8$ or $132/176$. This leads us to the following definition.

Definition 6.2. *Let $\cong$ be an equivalence relation on a set S. Let $a \in S$. We define the* **equivalence class** *of a to be*

$$[a] = \{x \in S \mid x \cong a\};$$

that is, $[a]$ is the set of all elements of S that are equivalent to a.

In Example 1, then, the equivalence class of a given triangle $\triangle ABC$ is just the set of all triangles that are *congruent* to $\triangle ABC$, whereas in Example 2 the equivalence class for $\triangle ABC$ is the much larger set of all triangles that are *similar* to $\triangle ABC$.

What do equivalence classes look like in Example 3? Let $a \in R$. Then an element $b \in R$ is in the equivalence class $[a]$ if and only if $b \cong a$ — that is, if and only if $b - a \in I$. In other words, $b \in [a]$ if and only if the cosets $b + I$ and $a + I$ are equal. We conclude that $[a] = a + I$. Thus, the equivalence classes are just the cosets.

In Example 4, the equivalence class $[3/4]$, for example, is the set of all fractions a/b such that $4a = 3b$ and $b \neq 0$, a condition that we normally write as $a/b = 3/4$. This is the sense in which the two fractions $3/4$ and $6/8$ are "equal": they are repesentatives of the same equivalence class. The key idea, then, is to *equate the equivalence class $[3/4]$ with a single rational number.*

With these examples in mind we can extract several basic ideas about equivalence classes.

The Basic Properties of Equivalence Classes

Let $\cong$ be an equivalence relation on a set S. Then the equivalence classes of S have the following basic properties:

1. each element $a \in S$ is in some equivalence class — namely, $[a]$;
2. $[a] = [b]$ if and only if $a \cong b$;
3. $[a] \cap [b] = \emptyset$ if and only if $a \not\cong b$;
4. two equivalence classes are either identical or disjoint;
5. the equivalence relation $\cong$ *partitions* the set S into equivalence classes.

Proofs

(1). Since $a \cong a$, $a \in [a]$.

(2). $[a] = [b]$ if and only if $a \in [b]$ and $b \in [a]$ if and only if $a \cong b$.

(3). Clearly, if $[a] \cap [b] = \emptyset$, then $a \not\cong b$. On the other hand, suppose $[a] \cap [b] \neq \emptyset$; then let $c \in [a] \cap [b]$. It follows that $c \cong a$ and $c \cong b$, but then by transitivity, we have $a \cong b$.

(4). This follows immediately from (2) and (3).

(5). This follows immediately from (1) and (4).

Let's clarify these ideas by looking at an example.

Example 5

Let $\cong$ be the equivalence relation defined on $\mathbf{Z}$ by

$$a \cong b \text{ if and only if } a \equiv b \pmod 5.$$

Then there are five equivalence classes — $[0]$, $[1]$, $[2]$, $[3]$, and $[4]$. These five sets are disjoint, and their union is the entire set $\mathbf{Z}$. Furthermore, any other equivalence class, such as $[1729]$, is identical to one of these five sets. Thus, this equivalence relation partitions $\mathbf{Z}$ into these five sets.

The Quotient Field of a Domain

We have seen how the rational numbers can be thought of merely as equivalence classes of fractions under a completely natural equivalence relation. We now describe this same procedure once more, but this time more precisely and in a slightly more general setting. Our goal is to create a field F beginning with an arbitrary integral domain D by imitating the process of creating the rationals $\mathbf{Q}$ from the integers $\mathbf{Z}$. This field F will be called the *quotient field* of the domain D. So, for example, $\mathbf{Q}$ will be the quotient field of $\mathbf{Z}$. (You need to be very careful with this terminology, and not confuse this concept with that of *quotient rings* discussed in Chapter 2. We seem to be stuck with these mildly confusing terms, but the context as well as the notation make it fairly easy to avoid misunderstandings.)

We begin by defining an equivalence relation on an integral domain D. Let D be an integral domain. We define an equivalence relation on the set S of "fractions" using elements of D,

$$S = \{a/b \mid a, b \in D, \ b \neq 0\},$$

by

$$\frac{a}{b} \cong \frac{c}{d} \text{ if and only if } ad = bc$$

for $a, b, c, d \in D, \ b, d \neq 0$.

We verify that this is in fact an equivalence relation:

1. reflexive — $a/b \cong a/b$, since $ab = ba$;
2. symmetric — $a/b \cong c/d$ means that $ad = bc$, and so $cb = da$, and $c/d \cong a/b$;
3. transitive — $a/b \cong c/d$, $c/d \cong e/f$ means that $ad = bc$, and so $adf = bcf$, but also $cf = de$, and so $bcf = bde$; we conclude that $adf = bde$, and so $d(af - be) = 0$. But D is an integral domain and $d \neq 0$, and so $af - be = 0$, which means that $a/b \cong e/f$.

This eqivalence relation on the set S of "fractions" from an integral domain determine a set F of equivalence classes of the form $[a/b]$, where $a, b \in D$ and $b \neq 0$. We wish to turn this set F of equivalence classes into a ring (in fact, F will be a field). In order to do this, we need to define two operations on the set F, addition and multiplication.

The natural definitions are

$$[a/b] + [c/d] = [(ad + bc)/bd]$$

and

$$[a/b] \cdot [c/d] = [ac/bd].$$

Note that the right-hand sides of these expressions make sense because D is a domain, and so $bd \neq 0$. Next, we must attend to the tedious details of showing that these operations are well defined — this means that the result of an addition or a multiplication does not depend upon the choice of the representatives for the equivalence classes being added or multiplied.

Suppose that $[a/b] = [a'/b']$ and $[c/d] = [c'/d']$. We must show that $[a/b] + [c/d] = [a'/b'] + [c'/d']$ and that $[a/b] \cdot [c/d] = [a'/b'] \cdot [c'/d']$. We know that $ab' = ba'$ and that $cd' = dc'$, so

$$(ad + bc)b'd' = adb'd' + bcb'd' = (ab')dd' + bb'(cd')$$
$$= (ba')dd' + bb'(dc') = bd(a'd' + b'c').$$

Thus, $[(ad + bc)/bd] = [(a'd' + b'c')/b'd']$, as desired, so addition is well defined.

Similarly,

$$acb'd' = (ab')(cd') = (ba')(dc') = bda'c',$$

and so $[ac/bd] = [a'c'/b'd']$, and multiplication is also well defined.

I now leave to you the routine details of showing that the set F together with these operations of addition and multiplication forms a ring (see Problem 6.2). The zero element of F is $[0/1]$ and the multiplicative identity is $[1/1]$. This ring F is in fact a field: if $[a/b] \in F$ and $[a/b] \neq [0/1]$, then $a \neq 0$, and so $[b/a] \in F$ and $[a/b]$ has an inverse in F — namely, $[b/a]$. This field F is important enough to warrant a special name:

Definition 6.3. *Let D be an integral domain. The above field F of equivalence classes of fractions from D, with addition and multiplication defined as above, is called the* **quotient field** *of D.*

Thus, the field **Q** of rational numbers is the quotient field of the ring **Z** of integers. It is standard practice to blur the distinction between a fraction and the equivalence class represented by that fraction. That is, we usually write a/b when we mean $[a/b]$. In addition, we even think of the domain D as being contained in its quotient field F, just as we think of **Z** as being contained in **Q**. We do this by identifying an element $a \in D$ with the fraction $a/1$ — that is, with $[a/1] \in F$.

The Total Quotient Ring

What we have succeeded in doing above is to show that we can always *embed* an integral domain in a field — that is, if D is an integral domain, then there exists a field F such that D is a subring of F. But, what if we begin with an arbitrary ring R? Can we always embed R in a field? One answer is: obviously not if R has zero-divisors, since a field has no zero-divisors, and so R could not be a subring of a field. This is exactly why we restricted the foregoing process to integral domains.

The main point of embedding a domain in a field is to *create* inverses for the elements of the domain — that is, to be able to divide by any nonzero element. This is what we achieved by creating the rationals from the integers. However, for an arbitrary ring we cannot hope to construct an inverse for a zero-divisor (why not?). But we can hope to construct inverses for any element that is *not* a zero-divisor. This is what we do next. First, we give a name to those elements in a ring that are not zero-divisors.

Definition 6.4. *Let R be a ring. An element a ∈ R is a* **regular element** *if it is not a zero-divisor.*

So, can we always divide by regular elements? Let's look at some examples.

Example 6
In particular, 0 is not a regular element in any nonzero ring. This is neither surprising nor troubling, since we don't expect to be able to divide by 0 anyway.

Example 7
The regular elements in $\mathbf{Z}/(12)$ are $1 + (12), 5 + (12), 7 + (12)$ and $11 + (12)$. Note that each of these elements already has an inverse within $\mathbf{Z}/(12)$.

Example 8
Consider the ring $\mathbf{Z} \times \mathbf{Z}$ — that is, all ordered pairs (a, b) with $a, b \in \mathbf{Z}$, and with addition and multiplication defined componentwise. Then an element (a, b) is regular if and only if $a \neq 0$ and $b \neq 0$. However, the only elements of $\mathbf{Z} \times \mathbf{Z}$ that have inverses are $(1, 1), (1, -1), (-1, 1)$, and $(-1, -1)$.

So, the answer is that you can't always divide by regular elements, but the very good news is that we will now show that any ring R can be *embedded* in a ring $Q(R)$ such that *every* regular element of R does have an inverse in $Q(R)$. An equivalence relation can be defined on an appropriate set S of "fractions" using elements of R,

$$S = \{a/b \mid a, b \in R, \text{ and } b \text{ is regular}\},$$

by

$$\frac{a}{b} \cong \frac{c}{d} \text{ if and only if } ad = bc$$

for all $a, b, c, d \in R$, where b and d are regular elements.

The proof that this does define an equivalence relation is almost exactly the same as that given above for integral domains, except that in this case you use the fact that the denominators are regular elements. This equivalence relation, then, determines a set $Q(R)$ of equivalence classes of the form $[a/b]$, where $a, b \in R$ and b is regular. We wish to turn this set $Q(R)$ of equivalence classes into a ring. In order to do this, we need to define two operations on the set $Q(R)$, addition and multiplication.

Again, the natural definitions are

$$[a/b] + [c/d] = [(ad + bc)/bd]$$

and

$$[a/b] \cdot [c/d] = [ac/bd].$$

Note that the right-hand sides of these expressions make sense because of the fact that the product of two regular elements b and d is also a regular element bd. The details of showing that the above definitions of addition and multiplication of equivalence classes are well defined are exactly the same as they were for the quotient field case. Similarly, the details of showing that the set $Q(R)$ together with the operations of addition and multiplication forms a ring are the same as before (see Problem 6.2).

Again, we emphasize that it is standard practice to ignore the distinction between a fraction and the equivalence class represented by that fraction. That is, we usually write a/b when we mean $[a/b]$. In particular, we think of the ring R as being contained in the ring $Q(R)$. We do this by identifying an element $a \in R$ with the quotient $a/1$ — that is, with $[a/1] \in Q(R)$. We have therefore succeeded in *embedding* R in a ring $Q(R)$ such that every regular element $b \in R$ has an inverse in $Q(R)$ — namely, $1/b$. Since the ring $Q(R)$ is apparently the ring of *all* possible quotients from R, we call $Q(R)$ the *total quotient ring* of R.

Definition 6.5. *Let R be a ring. The above ring $Q(R)$ of equivalence classes of fractions from R whose denominators are regular elements, with addition and multiplication defined as above, is called the **total quotient ring** of R.*

In the special case that the ring R is an integral domain, we shall continue to use the term quotient field. Once again, I remind you not to confuse the two very different concepts of *quotient ring*, Q/R, and *total quotient ring*, $Q(R)$.

Localization

It may seem that we have pushed this idea of constructing inverses for elements in a ring about as far as we can. Nonetheless, the construction of quotient fields and of total quotient rings are but special cases of a still more general procedure called *localization*. Localization represents one of the most fundamental techniques in commutative ring theory. In the case of a quotient field, we allow denominators to be any nonzero element; in the case of a total quotient ring we allow denominators to be any regular element; and, in the case of an arbitrary ring, we now allow denominators to be any element from what is called a *multiplicative system* (a concept you first encountered in Problem 3.3).

Definition 6.6. *Let R be a ring. A subset T of R is a* **multiplicative system** *if $1 \in T$, and if $a, b \in T$ implies that $ab \in T$ — that is, T is multiplicatively closed and contains* 1.

For example, in an integral domain the set of nonzero elements is a multiplicative system. More generally, in any ring the set of regular elements is a multiplicative system. One other example of a multiplicative system that is very important in commutative ring theory is the complement of a prime ideal (see Problem 3.3).

If T is a multiplicative system of a ring R, then an equivalence relation can be defined on an appropriate set S of "fractions" using elements of R,

$$S = \{a/b \mid a, b \in R, \text{ and } b \in T\},$$

by

$$\frac{a}{b} \cong \frac{c}{d} \text{ if and only if } t(ad - bc) = 0$$

for some $t \in T$.

We show that $\cong$ is in fact an equivalence relation on the set S of fractions whose denominators lie in the multiplicative system T. The relation $\cong$ is *reflexive*: $a/b \cong a/b$, since $1 \cdot (ab - ba) = 0$ for all $a, b \in R$, and $1 \in T$. The relation $\cong$ is *symmetric*: $a/b \cong c/d$ means that $t(ad - bc) = 0$ for some $t \in T$, and so $t(cb - da) = 0$, and $c/d \cong a/b$. The relation $\cong$ is *transitive*: $a/b \cong c/d, c/d \cong e/f$ means that $s(ad - bc) = 0$ for some $s \in T$, and $t(cf - de) = 0$ for some $t \in T$; multiplying the first equation by ft and the second by bs yields

$$ftsad - ftsbc = 0 \text{ and } bstcf - bstde = 0;$$

adding these two equations gives us $ftsad - bstde = 0$; thus, $dst(af - be) = 0$; since T is multiplicatively closed, $dst \in T$; and we conclude that $a/b \cong e/f$.

This equivalence relation on the set S of fractions from R whose denominators are elements of the multiplicative system T determines a set R_T of equivalence classes of the form $[a/b]$, where $a, b \in R$, and $b \in T$. We wish to turn this set R_T of equivalence classes into a ring. In order to do this, we need to define two operations on the set R_T, addition and multiplication.

Once again, the natural definitions are

$$[a/b] + [c/d] = [(ad + bc)/bd]$$

and

$$[a/b] \cdot [c/d] = [ac/bd].$$

Note that the right-hand sides of these expressions make sense because T is multiplicatively closed, and so $bd \in T$. The details of showing that the above definitions of addition and multiplication of equivalence classes are well defined are left to you (see Problem 6.6), as well as the details of showing that the set R_T together with the operations of addition and multiplication forms a ring (see Problem 6.7). The zero element of R_T is $[0/1]$ and the multiplicative identity is $[1/1]$. It is standard practice to ignore the distinction between a fraction and the equivalence class represented by that fraction — that is, we normally write a/b instead of $[a/b]$.

Definition 6.7. *Let R be a ring, and let T be a multiplicative system of R. The above ring R_T of equivalence classes of fractions from R whose denominators are elements of T, with addition and multiplication defined as above, is called the* **localization** *of R with respect to T, and which we read as "R localized at T."*

Example 9

Let T be the set of *odd* integers in $\mathbf{Z}$. Then T is a multiplicative system in $\mathbf{Z}$. The localization $\mathbf{Z}_T$ consists of all fractions a/b such that b is odd. Note that $a/b \cong c/d$ if and only if $t(ad - bc) = 0$ for some odd integer T, but $\mathbf{Z}$ is an integral domain, so this condition is equivalent to $ad - bc = 0$, which is just the usual condition for two fractions to be equal. Thus, $\mathbf{Z}_T$ is the subring of the rationals $\mathbf{Q}$ consisting of fractions having odd denominators.

Example 10

Let p be a prime number. Let T be the set of integers in $\mathbf{Z}$ that are relatively prime to p. Then T is a multiplicative system in $\mathbf{Z}$. The localization $\mathbf{Z}_T$ consists of all rational numbers a/b such that b is relatively prime to p.

Example 11

We can generalize these last two examples. Let P be a prime ideal of a ring R. Let $T = R\backslash P$. Then T is a multiplicative system. The localization R_T consists of all fractions a/b, such that b is *not* in the ideal P. This situation is common enough that we have a special notation: rather than use $R_{R\backslash P}$, which is correct but awkward, we use instead R_P. Thus, the rings of Examples 9 and 10 are written, respectively, as $\mathbf{Z}_{(2)}$ and $\mathbf{Z}_{(p)}$. You need to be careful to distinguish between the "ring of fractions," written as $\mathbf{Z}_{(p)}$, and the "integers mod p," written as $\mathbf{Z}/(p)$. You should also be aware that not everyone uses this same notation.

One of the reasons that localization is such a useful process in commutative ring theory is that there is a very pretty relationship between the ideals of R and the ideals of the localization R_T. Because of the importance of this relationship, we will pursue it in some detail. If you find these details overly tedious, you should move on immediately to the statement of this relationship given in the next theorem. I assure you, your loss will be minimal.

Let R be a ring, and let T be a multiplicative system of R. Our objective is to find a correspondence between the set of ideals of R and the set of ideals of R_T. With this in mind, for an ideal I of R, we define a set I_T in R_T by

$$I_T = \{x/t \mid x \in I, t \in T\}.$$

Then, we claim that the set I_T is an ideal of the ring R_T. Let $x/s, y/t \in I_T$ — that is, $x, y \in I, s, t \in T$; then

$$\frac{x}{s} - \frac{y}{t} = \frac{xt - sy}{st} \in I_T,$$

since I is an ideal and T is multiplicatively closed. Similarly, let $a/s \in R_T$, and $x/t \in I_T$; then

$$\frac{a}{s} \cdot \frac{x}{t} = \frac{ax}{st} \in I_T;$$

thus I_T is an ideal of R_T, as claimed. Therefore, for any ideal I of the ring R we can associate with I — in a very natural way — an ideal I_T of the ring R_T. We now show that we can also play this game in the other direction.

Let J be an ideal of R_T. (In particular, recall that this means that the elements of J are fractions.) For this ideal J, we define a subset I of R by

$$I = \{x \in R \mid x/1 \in J\}.$$

Then we claim that the set I is an ideal of the ring R. Let $x, y \in I$; then $x/1, y/1 \in J$, so

$$\frac{x-y}{1} = \frac{x}{1} - \frac{y}{1} \in J,$$

since J is an ideal; thus, $x - y \in I$. Similarly, let $a \in R$, and $x \in I$; then $a/1 \in R_T$ and $x/1 \in J$, so

$$\frac{ax}{1} = \frac{a}{1} \cdot \frac{x}{1} \in J,$$

since J is an ideal; thus, $ax \in I$. We conclude that I is an ideal of R, as claimed. Therefore, for any ideal J of the ring R_T we can associate with J — again, in a very natural way — an ideal I of the ring R.

We can summarize the discussion thus far by saying that we have successfully defined two functions. The first is a function Φ:

$$\Phi : \text{ the set of ideals of } R \longrightarrow \text{ the set of ideals of } R_T$$

defined by

$$\Phi : I \longmapsto I_T.$$

The second is a function Ψ in the other direction:

$$\Psi : \text{ the set of ideals of } R_T \longrightarrow \text{ the set of ideals of } R$$

defined by

$$\Psi : J \longmapsto I,$$

where $I = \{x \in R \mid x/1 \in J\}$.

There is a French saying that there are two kinds of questions: *une question qu'on se pose—une question qui se pose*; that is, questions that one poses and questions that pose themselves. Clearly, we are now faced with an obvious question that poses itself: what happens if we begin with an ideal I in R, and carry this ideal to an ideal I_T in R_T, and then back again to an ideal in R in the natural fashion described above? Do we get back to the ideal I we started with in the first place, or not? Similarly, we could ask the same question beginning with an ideal J in R_T. In other words, is it always true that $\Psi(\Phi(I)) = I$, and that $\Phi(\Psi(J)) = J$?

Happily, we can easily show that starting in one direction everything is as it should be. That is, if we let J be an ideal in R_T, then it is in fact true that $\Phi(\Psi(J)) = J$. Let $I = \Psi(J)$ — that is, $I = \{x \in R \mid x/1 \in J\}$. Since $\Phi(I) = I_T$, we must show that $I_T = J$. In order to show that these two sets are equal, we will show each is contained in the other.

First, let $x/t \in I_T$, where $x \in I$ and $t \in T$. We must show that $x/t \in J$. But $x \in I$ means that $x/1 \in J$, and so, since $1/t \in R_T$ and J is an ideal of R_T,

$$\frac{x}{t} = \frac{1}{t} \cdot \frac{x}{1} \in J,$$

as desired. Thus $I_T \subseteq J$.

Now let $x/t \in J$, where $x \in R$ and $t \in T$. We must show that $x/t \in I_T$ — that is, we must show that $x \in I$. But, $t/1 \in R_T$ and J is an ideal of R_T, so

$$\frac{x}{1} = \frac{t}{1} \cdot \frac{x}{t} \in J,$$

which means that $x \in I$. Therefore, $J \subseteq I_T$. Thus, $I_T = J$, and so it is true that $\Phi(\Psi(J)) = J$.

We now come to some very sad news: it is *not* always true that $\Psi(\Phi(I)) = I$, as we see in the next example:

Example 12
We saw in Example 9 that $\mathbf{Z}_{(2)}$, the ring of rational numbers with odd denominators, is the localization of $\mathbf{Z}$ with respect to T, where $T = \mathbf{Z} \setminus (2)$ is the multiplicative system consisting of the odd integers. We take $I = (3)$ for our example. What is the ideal I_T? It is simply all fractions x/t such that $x \in I$ and $t \in T$ — that is, x is divisible by 3 and t is odd. Now, what is $\Psi(I_T)$? By definition, it is the set

$$I' = \{x \in \mathbf{Z} \mid x/1 \in I_T\}.$$

But we claim that I' is the *whole* of $\mathbf{Z}$ (rather than just I, as we had hoped). We show that any element of $\mathbf{Z}$ is in I'. Let $n \in \mathbf{Z}$. Then $3n \in I$, since I is an ideal of $\mathbf{Z}$, and so $n/1 = 3n/3 \in I_T$. Therefore, $n \in I'$. Thus, $I' = \mathbf{Z}$. There is a picturesque way to decribe this unfortunate situation: we began with an ideal (3) in $\mathbf{Z}$, which we took over to the ring $\mathbf{Z}_{(2)}$, but when we brought it back it had *blown up* to be the *entire* ring $\mathbf{Z}$.

At this point we need to regroup. The correspondence between the set of ideals of R and those of R_T turns out not to be as nice as we had hoped. The question *we* now pose is: can we place some restrictions on the situation so that the correspondence *is* as nice as we originally had hoped? That the answer to this question is *yes* is the substance of our next theorem.

Theorem 6.1. *Let R be a ring, and let T be a multiplicative system of R. There is a one-to-one order-preserving correspondence between the prime ideals of the localization R_T and the prime ideals of R which are disjoint from T.*

Proof

The correspondence is given by the function

$$\Phi : P \longmapsto P_T$$

for any prime ideal P of the ring R that is disjoint from T. There are several routine matters to check.

First, note that P_T is a proper ideal of R_T, since $1/1 \notin P_T$, for if $1/1 = p/t \in P_T$, where $p \in P$ and $t \in T$, then $s(t - p) = 0$, for some $s \in T$, but $s \notin P$, so $t - p \in P$, and $t \in P$, which is a contradiction. Thus, P_T is a proper ideal of R_T.

Next, we must make sure that P_T is a prime ideal. Suppose that $(x/s)(y/t) \in P_T$, where $x, y \in P$ and $s, t \in T$. We must show that $x/s \in P_T$ or $y/t \in P_T$. Since $(x/s)(y/t) \in P_T$, we can write $xy/st = p/r$, for some $p \in P, r \in T$. Then, there is an element $w \in T$ such that $w(xyr - stp) = 0$. Now, $wstp \in P$ and $0 \in P$, so it follows that $wrxy \in P$. But, $wr \notin P$, since P is disjoint from T, so $xy \in P$. Hence, $x \in P$ or $y \in P$, and so $x/s \in P_T$ or $y/t \in P_T$, and we conclude that P_T is a prime ideal.

Next, we observe from the definition of I_T that the function Φ is order-preserving — that is, if $I \subseteq J$, then $\Phi(I) \subseteq \Phi(J)$. So it remains only to verify that the correspondence Φ is *one-to-one* and *onto*.

One-to-one: Suppose that $P_T = Q_T$ for two prime ideals P and Q, each disjoint from T. We must show that $P = Q$. By symmetry, it is enough to show that $P \subseteq Q$. Let $p \in P$. Then $p/1 \in P_T$, so $p/1 \in Q_T$. Thus, $p/1 = q/s$ for some $q \in Q$ and $s \in T$. This means that there is an element $t \in T$ such that $t(ps - q) = 0$. Hence, $tps \in Q$, from which it follows that $p \in Q$, since $t \notin Q$ and $s \notin Q$. Thus, $P \subseteq Q$, as desired, and Φ is one-to-one.

Onto: Let Q be a prime ideal of R_T. We must find a *preimage* of Q — that is, we must find a prime ideal P of R disjoint from T such that $P_T = Q$. The likely candidate is $P = \{x \in R \mid x/1 \in Q\}$. We must show four things: that P is a proper ideal, that P is a prime ideal, that P is disjoint from T, and that $P_T = Q$. P is a proper ideal, since if $1 \in P$, then $1/1 \in Q$, and this contradicts the fact that Q is a proper ideal of R_T. Next, we show that P is prime. Let $xy \in P$. We must show that $x \in P$ or $y \in P$. Now, $xy \in P$ means that $xy/1 \in Q$; but $xy/1 = (x/1)(y/1)$, so $(x/1)(y/1) \in Q$. Since Q is prime, $x/1 \in Q$ or $y/1 \in Q$, so $x \in P$ or $y \in P$. Thus P is a prime ideal. Next, we show that P is disjoint from T. If $t \in P \cap T$, then $t/1 \in Q$, and so $1 = (1/t)(t/1) \in Q$. Thus, $Q = R_T$, and hence Q is not a proper ideal of R_T. This is a contradiction, since Q is a prime ideal of R_T. Hence, $P \cap T = \emptyset$. Finally, the fact that $P_T = Q$ follows from the general fact that $\Phi(\Psi(J)) = J$, which we proved above. This completes the proof of the theorem.

Example 13

Let us consider once again the ring $\mathbf{Z}$ and the localization $\mathbf{Z}_{(2)}$ consisting of all rational numbers with odd denominators. Recall that, in this case, the multiplicative system is the set T consisting of the odd integers. What exactly does Theorem 6.1 tell us in this situation? Let P be a nonzero prime ideal of $\mathbf{Z}$ disjoint from T. Then $P = (p)$, where p is a prime number and the ideal (p) contains no odd numbers. Hence, p is not an odd prime, so $p = 2$. Thus, the *only* nonzero prime ideal disjoint from T is (2). The theorem then tells us that correspondingly the ring $\mathbf{Z}_{(2)}$ has only *one* nonzero prime ideal — namely, the ideal $(2)_T$ consisting of all rational numbers a/b, where a is even and b is odd. Note that this ideal, being the only nonzero prime ideal of $\mathbf{Z}_{(2)}$, is the *unique maximal ideal* of the ring $\mathbf{Z}_{(2)}$.

Example 14

Let us return to Example 11, where P is a prime ideal of a ring R and the localization R_P is the ring of all fractions a/b such that $a \in R$ and $b \notin P$. In this case we say we have *localized at P*. What does Theorem 6.1 tell us in this situation? It says there is a one-to-one order-preserving correspondence between the proper ideals of R_P and the prime ideals of R which are *contained* in P. Since P is obviously the largest prime ideal contained in P, this means that the ideal of R_P that corresponds to P is the *unique maximal ideal* of R_P. Thus, the unique maximal ideal of R_P is

$$P_P = \{x/t \mid x \in P, t \notin P\}.$$

We use this important example to motivate the following definition:

Definition 6.8. *A ring is said to be a* **local** *ring if it has a unique maximal ideal.*

Problems

6.1   Let R be a set and let S be a *partition* of R — that is, $S = \{S_\lambda\}_{\lambda \in \Lambda}$, over an index set Λ such that

      1. $S_\lambda \subseteq R$, for each $\lambda \in \Lambda$;

      2. $R = \bigcup_{\lambda \in \Lambda} S_\lambda$;

      3. $S_\alpha \cap S_\beta = \emptyset$, for all $\alpha \neq \beta \in \Lambda$.

    We can define a relation $\cong$ on R as follows:

$$a \cong b \text{ if and only if } a, b \in S_\lambda, \text{ for some } \lambda \in \Lambda.$$

Show that $\cong$ is an equivalence relation on R. What are the equivalence classes?

6.2   Let D be an integral domain. Verify the details of showing that the set F of equivalence classes of fractions from D together with the operations of addition and multiplication of equivalence classes forms a ring. How does this proof change when verifying the details of showing that the total quotient ring is a ring?

6.3   Find the total quotient rings $Q(\mathbf{Z})$ and $Q(\mathbf{Z} \times \mathbf{Z})$.

6.4   Find the total quotient ring $Q(\mathbf{Z}/(12))$.

6.5   In the process of localizing a ring R with respect to a multiplicative system T, we defined an equivalence relation $\cong$ as follows:

$$a/b \cong c/d \text{ if and only if } t(ad - bc) = 0,$$

for some $t \in T$.

    Suppose instead we had continued as before with total quotient rings and defined the relation $\cong$ as follows:

$$a/b \cong c/d \text{ if and only if } ad - bc = 0.$$

Show that this does not in general give us an equivalence relation because transitivity need not hold. In particular, let R be the ring $\mathbf{Z}/(12)$, and let T be the multiplicative system $\{1 + (12), 2 + (12), 4 + (12), 8 + (12)\}$. Then find three fractions with numerators in R and denominators in T for which transitivity fails.

6.6   Let R be a ring and let T be a multiplicative system of R. Prove that the operations of addition and multiplication of equivalence classes in the localization R_T are well defined.

6.7   Let R be a ring and let T be a multiplicative system of R. Verify in detail that the localization R_T is a ring.

6.8   Explain how both the quotient field of an integral domain and the total quotient ring of an arbitrary ring can be thought of as localizations.

6.9   Let R be a ring and let T be a multiplicative system of R. Prove that

$$R_T = 0 \text{ if and only if } 0 \in T.$$

It is therefore not very interesting for us to let a multiplicative system contain 0.

6.10   Let R be a ring and let T be the set of all *units* of R. Describe the ring R_T.

6.11   Let $n \neq 0 \in \mathbf{Z}$. Let $T = \{n^k \mid k \geq 0\}$ — that is, let T consist of all the powers $1, n, n^2, \ldots$ of n. Show that T is a multiplicative system of $\mathbf{Z}$. Describe the elements of the ring $\mathbf{Z}_T$.
        What property would an element a of an arbitrary ring R have to have in order that we could similarly define the localization of R with respect to the powers of a — that is, in order that we wouldn't have to worry that the localization could just turn out to be the zero ring?

6.12   Let R be the ring $\mathbf{Z}/(12)$ and let a be the element $2 + (12)$. Then let $T = \{a^k \mid k \geq 0\}$ — that is,

$$T = \{1 + (12), 2 + (12), 4 + (12), 8 + (12)\}.$$

Describe the elements of the ring R_T. (Hint: there are only three!)

6.13   Consider the ring $K[x]$ of polynomials over a field K. Let $a \in K$. Then, let T be all polynomials $f \in K[x]$ such that $f(a) \neq 0$. Show that T is a multiplicative system of $K[x]$. Describe the elements of $K[x]_T$.

6.14   Prove that *every* ring between $\mathbf{Z}$ and $\mathbf{Q}$ is a localization of $\mathbf{Z}$ — that is, prove that if R is a ring such that

$$\mathbf{Z} \subseteq R \subseteq \mathbf{Q},$$

then there is a multiplicative system T of $\mathbf{Z}$ such that $R = \mathbf{Z}_T$. (Hint: the first thing to do is to figure out what the set T should be.)

6.15   In this problem we shall see that two different multiplicative systems in a ring R can give rise to the same localization. Find two multiplicative systems T_1 and T_2 in $\mathbf{Z}$ such that the localizations $\mathbf{Z}_{T_1}$ and $\mathbf{Z}_{T_2}$ are the same.

6.16   We saw in Example 12 that if I is an arbitrary ideal of a ring R, then $\Psi(\Phi(I))$ need not be I — in fact, it could be the entire ring R. Show, more generally, that this same thing happens whenever I is an ideal that is *not* disjoint from the multiplicative system T of a ring R.

6.17   Let p be a prime number. Show that $\mathbf{Z}_{(p)}$ is a local ring. What is its unique maximal ideal?

6.18   Let P be a prime ideal of a ring R. In Example 14 we used Theorem 6.1 to show that R_P is a local ring. Prove directly that R_P is a local ring by showing that the set

$$M = \{x/t \mid x \in P, t \notin P\}$$

is an ideal consisting of all non-units of R_T, and hence, is the unique maximal ideal of R_T.

6.19   Let $\mathcal{N}$ be the nilradical of a ring R. Let T be a multiplicative system of R. Show that $\mathcal{N}_T$ is the nilradical of R_T.

6.20   A ring-theoretic property $\mathcal{P}$ is called a *local property* if a ring R has property $\mathcal{P}$ if and only if each localization R_P of the ring at a prime ideal P also has property $\mathcal{P}$. The idea, then, is to be able to show that a ring has a certain property by showing that it has the property locally.

   Show that the property of not having any non-zero nilpotent elements is a local property — that is, show that a ring R has no non-zero nilpotent elements if and only if R_P has no non-zero nilpotent elements for each prime P.

   Is the property of being an integral domain a local property?

7

Rings of Continuous Functions

In this chapter, we turn our attention to a specific commutative ring and begin to explore the profound relationship between commutative ring theory and an entirely different branch of mathematics called topology. For the time being, we will restrict our attention to the topology of the real line. In particular, we shall be interested in real-valued functions defined on the closed interval $[0, 1]$ — that is, functions of the form

$$f : [0, 1] \longrightarrow \mathbf{R},$$

where $\mathbf{R}$ is the set of real numbers.

The set of *all* such functions $f : [0, 1] \to \mathbf{R}$ will be denoted by $\mathbf{R}^{[0,1]}$. We use this notation — the *power* notation — because it reflects the fact that the number m^n — that is, m raised to the nth power — is precisely the number of different functions

$$f : N \longrightarrow M,$$

where $N = \{1, 2, \ldots, n\}$ and $M = \{1, 2, \ldots, m\}$. This is because in order to define a function f from N to M there are m choices to be made for each image $f(a)$, and a choice has to be made for each of the n different elements a in N; thus, there are m^n choices in all. In general, then, we use the power notation Y^X to represent the set of *all* functions $f : X \to Y$.

The set $\mathbf{R}^{[0,1]}$ of real-valued functions defined on the closed interval $[0, 1]$ is a ring in which the operations of addition and multiplication of these functions are defined by

$$(f + g)(x) = f(x) + g(x) \quad \text{and} \quad (fg)(x) = f(x)g(x).$$

In other words, addition and multiplication are defined *pointwise*. Since each of these operations is defined pointwise within the ring of real numbers — that is, the addition $f(x) + g(x)$ and the multiplication $f(x)g(x)$ take place in the ring $\mathbf{R}$ — it is easy to verify that the associative, distributive, and commutative properties for rings hold in $\mathbf{R}^{[0,1]}$ because each of these properties holds in the ring $\mathbf{R}$.

The zero element of the ring $\mathbf{R}^{[0,1]}$ is the function which is identically zero — that is, the function f such that $f(x) = 0$ for all $x \in [0, 1]$. As is customary for the zero element of a ring, we use the symbol 0 to denote this function; thus, $0(x) = 0$ for all $x \in [0, 1]$. The multiplicative identity is the function which is identically 1 — that is, the function f such that $f(x) = 1$ for all $x \in [0, 1]$. We denote this function by the symbol 1; thus, $1(x) = 1$ for all $x \in [0, 1]$.

What are the units in the ring $\mathbf{R}^{[0,1]}$? A function f is a unit in $\mathbf{R}^{[0,1]}$ if and only if there is a function g such that $fg = 1$ — that is, a function g such that $f(x)g(x) = 1$ for all $x \in [0, 1]$. This means that $f(x) \neq 0$ for *all* $x \in [0, 1]$. So the units in $\mathbf{R}^{[0,1]}$ are just the functions that are never zero — that is, the functions whose graphs never touch the x-axis. In this case, the inverse of a unit f is the function g given by $g(x) = \frac{1}{f(x)}$. We sometimes denote this inverse by $\frac{1}{f}$.

The ring $\mathbf{R}^{[0,1]}$ is certainly not an integral domain and, hence, it is also not a field. For example, let $f, g \in \mathbf{R}^{[0,1]}$ be two functions defined by

$$f(x) = \begin{cases} 0 & \text{for } 0 \leq x \leq \frac{1}{2}, \\ 1 & \text{for } \frac{1}{2} < x \leq 1 \end{cases}$$

and

$$g(x) = \begin{cases} 1 & \text{for } 0 \leq x \leq \frac{1}{2}, \\ 0 & \text{for } \frac{1}{2} < x \leq 1. \end{cases}$$

Then $fg = 0$, but neither f nor g is 0. (Make sure you recognize in this paragraph when the symbol "0" is being used to represent the number 0 and when it is being used to represent the function 0.)

However, it is not the ring $\mathbf{R}^{[0,1]}$ that holds our primary interest in this chapter; it is rather the subring of $\mathbf{R}^{[0,1]}$ consisting of those functions that are *continuous* that will be our main focus. The continuous real-valued functions defined on the closed interval [0, 1] do in fact form a subring of $\mathbf{R}^{[0,1]}$: the sum or product of two continuous functions is again a continuous function; also, if f is a continuous function, then so is its additive inverse $-f$; and, finally, the multiplicative identity 1 is a continuous function. We call this subring of $\mathbf{R}^{[0,1]}$ the *ring of continuous functions* from [0, 1] to $\mathbf{R}$, and denote this subring by $C([0, 1])$, or more simply just as C, keeping in mind that the domain for these continuous real-valued functions is the topological space [0, 1].

Our rather modest goal in this chapter, then, will be to describe the maximal ideals of the ring $C([0, 1])$. First of all, we know by Theorem 5.1 that *no* unit can lie in any maximal ideal, but that any other element must be in *some* maximal ideal. But, the unit elements in $C([0, 1])$ are just the functions that are never 0 — that is, functions whose graphs never touch or cross the x-axis. Therefore, if a function f is 0 for some point $p \in [0, 1]$ — that is, if f touches or crosses the x-axis at p — then f is not a unit, and so f must be contained in some maximal ideal. This simple idea will allow us to get started by describing at least some of the maximal ideals of $C([0, 1])$. Whether this idea will allow us to describe *all* the maximal ideals remains to be seen.

For any point $p \in [0, 1]$, let M_p denote the set of all functions in the ring $C([0, 1])$ that are 0 at p — that is, all functions f such that $f(p) = 0$. Thus, for $p \in [0, 1]$, we define

$$M_p = \{ f \in C([0, 1]) \mid f(p) = 0 \}.$$

You should visualize the elements of M_p as continuous functions whose graphs must pass through the point $(p, 0)$ — that is, they "touch" or "cross" the x-axis at p. As we shall show, the set M_p is a maximal ideal of the ring $C([0, 1])$.

First, we show that the set M_p is in fact an ideal of the ring $C([0, 1])$. If $f, g \in M_p$, then $f(p) = 0$ and $g(p) = 0$, and so $(f - g)(p) = f(p) - g(p) = 0 - 0 = 0$, and so $f - g \in M_p$; if $f \in C([0, 1])$ and $g \in M_p$; then $g(p) = 0$, so $(fg)(p) = f(p)g(p) = f(p) \cdot 0 = 0$, and so $fg \in M_p$. Thus M_p is an ideal. Moreover, M_p is a *proper* ideal since, for example, the function 1 is not 0 at p, that is, $1 \notin M_p$.

Now, in order to show that M_p is actually a *maximal* ideal of $C([0, 1])$, we show that if N is an ideal such that $M_p \subset N$, then $N = C([0, 1])$. So, let $f \in N$ such that $f \notin M_p$. We will show that $(M_p, f) = C([0, 1])$, from which it follows that $N = C([0, 1])$. (Recall from page 29 in Chapter 3 that the notation (I, a) means the ideal generated by I and a.) In order to show that the ideal (M_p, f) is the entire ring $C([0, 1])$, we will produce a *unit* in (M_p, f).

Let g be an element of $C([0, 1])$ that is 0 *only* at p — that is, $g(p) = 0$, but $g(x) \neq 0$, for $x \neq p$. In other words, g touches or crosses the x-axis only at p. (For example, we could let g be the function $g(x) = x - p$.) In particular, $g \in M_p$, so the function $g^2 + f^2$ is contained in the ideal (M_p, f). We claim that $g^2 + f^2$ is a unit. Since $f \notin M_p$, the function f is not 0 at p, whereas the function g is 0 *only* at p. Therefore, the function $g^2 + f^2$ is not 0 anywhere in the interval $[0, 1]$, and so $g^2 + f^2$ is a unit. Moreover, $g^2 + f^2$ is not only a unit, but is also in the ideal (M_p, f). We conclude that (M_p, f) must be the entire ring $C([0, 1])$. Hence, as claimed, M_p is a maximal ideal for each $p \in [0, 1]$.

We have just seen that for each point p in $[0, 1]$ there is a corresponding maximal ideal of $C([0, 1])$ — namely, M_p. In other words, we have already found lots of maximal ideals in the ring $C[0, 1])$. We now show that these maximal ideals M_p are the *only* maximal ideals in the ring $C([0, 1])$! Not surprisingly, this will take a bit more effort.

Let M be a maximal ideal of $C([0, 1])$. We must show that M is of the form M_p — that is, we must show that $M = M_p$ for some point $p \in [0, 1]$. Our goal, then, is to produce such a point p. In order to do this, we need to know a few basic facts about the topology of the closed interval $[0, 1]$. We therefore temporarily suspend our discussion of maximal ideals in $C([0, 1])$ in order to make several necessary remarks about topological matters.

Specifically, what we shall be interested in is the behavior of various infinite collections of intervals of real numbers, that is, collections of the form

$$\mathcal{I} = \{I_1, I_2, I_3, \dots\},$$

where each of the sets I_n is an interval. In particular, we shall be interested in the behavior of these intervals when they are *nested*, that is, each interval in the sequence is contained in the previous interval:

$$I_1 \supseteq I_2 \supseteq I_3 \supseteq \cdots .$$

The main question we shall be concerned with in this situation is the following:

what is the *intersection* of a such nested set of intervals?

In other words, what can we say about the intersection

$$\bigcap_{n \geq 1} I_n \, ?$$

Can the intersection be empty, or must there always be at least one point that is in every interval?

Since we can be badly fooled by our intuition on matters such as these, let us look at some examples. In order to make the general facts about such intersections as clear as possible we will use the real line for our examples. In the end, however, we shall return our attention to the set $[0, 1]$.

Example 1
Let $\mathcal{I} = \left\{ (0, \frac{1}{n}) \right\}_{n \geq 1}$ — that is, $\mathcal{I}$ is the nested sequence of *open* intervals

$$(0, 1) \supset (0, \tfrac{1}{2}) \supset (0, \tfrac{1}{3}) \supset (0, \tfrac{1}{4}) \supset (0, \tfrac{1}{5}) \supset \cdots .$$

I hope it is obvious to you that the intersection of these intervals is empty. There can be no positive real number that is in *all* these sets, simply because whatever positive real number you think might work, eventually $\frac{1}{n}$ becomes smaller than this number, and so this number is not in the set $(0, \frac{1}{n})$ after all.

Example 2

Let $\mathcal{I} = \{[0, \frac{1}{n}]\}_{n \geq 1}$ — that is, $\mathcal{I}$ is the nested sequence of *closed* intervals

$$[0, 1] \supset [0, \tfrac{1}{2}] \supset [0, \tfrac{1}{3}] \supset [0, \tfrac{1}{4}] \supset [0, \tfrac{1}{5}] \supset \cdots .$$

The intersection of these intervals is not empty, of course, since the number 0 is in each interval. In fact, the intersection consists of just the single number 0.

It might appear from these two examples as if a sequence of nested *open* intervals has an empty intersection, whereas a sequence of nested *closed* intervals has a non-empty intersection. The next examples show us that things aren't quite that simple.

Example 3

Let $\mathcal{I} = \{(\frac{n}{n+1}, 2)\}_{n \geq 1}$ — that is, $\mathcal{I}$ is the nested sequence of *open* intervals

$$(\tfrac{1}{2}, 2) \supset (\tfrac{2}{3}, 2) \supset (\tfrac{3}{4}, 2) \supset (\tfrac{4}{5}, 2) \supset (\tfrac{5}{6}, 2) \supset \cdots .$$

Clearly, the intersection of these intervals is not empty. In fact, the intersection is the half-open, half-closed interval $[1, 2)$.

Example 4

Let $\mathcal{I} = \{[n, \infty)\}_{n \geq 1}$ — that is, $\mathcal{I}$ is the nested sequence of *closed* intervals

$$[1, \infty) \supset [2, \infty) \supset [3, \infty) \supset [4, \infty) \supset [5, \infty) \supset \cdots .$$

Obviously, the intersection of these intervals is empty. There is no number that is in *all* these sets, again because whatever number you think of, eventually n becomes larger than this number, and so the number is not in the set $[n, \infty)$ after all.

With these four examples as background, we can now state the critical topological property that we need to know about the closed interval $[0, 1]$ in order to resume our discussion of the maximal ideals in $C([0, 1])$. Let

$$I_1 \supseteq I_2 \supseteq I_3 \supseteq \cdots$$

be a set of nested *closed* intervals in the topological space $[0, 1]$; then

the intersection of these intervals is non-empty.

In other words, there is at least one point p such that p lies in every interval.

If, further, the lengths of the intervals I_n approach 0 as $n \to \infty$, then

there is a single point in the intersection of these intervals.

This last property is intuitively evident. For example, it is this intersection property of the real line that lies behind our firm belief that the number $3.14159...$ represents a *unique* point on the real line — namely, the single point π that lies in the intersection of the nested sequence of closed intervals

$$[3, 4] \supset [3.1, 3.2] \supset [3.14, 3.15] \supset [3.141, 3.142] \supset \cdots .$$

Returning now to the main task at hand, recall that M is a maximal ideal of $C([0, 1])$ and we are trying to represent M as an ideal of the form M_p — that is, we are trying to find a number $p \in [0, 1]$ such that $M = M_p$. It is the above-mentioned intersection property of the interval $[0, 1]$ that will produce the number p for us. We just have to find an appropriate sequence of nested closed intervals in order to produce p.

Let n be a positive integer. We can then easily find n functions $h_1, h_2, \ldots, h_n$ in $C([0, 1])$ with the property that, for each k such that $1 \leq k \leq n$, the function h_k is 0 on the interval $[\frac{k-1}{n}, \frac{k}{n}]$ and is positive everywhere else. For example, if $n = 3$, then the function h_1 must be 0 on the interval $[0, \frac{1}{3}]$ and be positive elsewhere, the function h_2 must be 0 on the interval $[\frac{1}{3}, \frac{2}{3}]$ and be positive elsewhere, and the function h_3 must be 0 on the interval $[\frac{2}{3}, 1]$ and be positive elsewhere. You could easily draw three such functions.

By the way in which the functions $h_1, h_2, \ldots, h_n$ were chosen, we see that the product of these n functions is the zero function, and hence, this product is in the ideal M — that is,

$$h_1 h_2 h_3 \cdots h_n = 0 \in M.$$

But M is a prime ideal (since it is a maximal ideal), and so one of the functions h_k lies in M. We *rename* this particular function g_n.

In this way, for each n, we can find a function g_n in M that is 0 only on a closed interval of length $\frac{1}{n}$ — that is, we now have an infinite sequence of functions

$$g_1, g_2, g_3, \cdots$$

in M such that the function g_1 is 0 only on a closed interval of length 1, the function g_2 is 0 only on a closed interval of length $\frac{1}{2}$, the function g_3 is 0 only on a closed interval of length $\frac{1}{3}$, and so on. However, since these particular intervals might not be nested, we need to modify this particular sequence of functions and create yet another sequence of functions, still in M, that have *nested* closed intervals on which they are 0.

To this end we define the following functions:

$$f_1 = g_1, \ f_2 = f_1 + g_2, \ f_3 = f_2 + g_3, \ f_4 = f_3 + g_4, \text{ and so on.}$$

Then, obviously, $f_n \in M$ for each n since $g_n \in M$ for each n. Next, note that, for each $n > 1$, f_n is 0 only at points where *both* f_{n-1} and g_n are 0. This is because all the functions h_k were positive except on the interval where they were 0, and so all the functions g_n, and thus all the functions f_n, also are positive except where they are 0. Hence, the functions $f_1, f_2, f_3, \ldots$ are elements of M such that f_1 is 0 on a closed interval I_1, f_2 is 0 on a closed interval I_2, and so on, giving us a set $\mathcal{I}$ of *nested* closed intervals I_n

$$I_1 \supset I_2 \supset I_3 \supset \cdots$$

whose lengths, $\frac{1}{n}$, approach 0 as $n \to \infty$. Let p be the *single* point in the intersection of these nested closed intervals. We shall show that $M = M_p$.

Let $f \in M$. Then $f^2 + f_n \in M$, for each n. So, for a fixed n, $f^2 + f_n$ is not a unit in $C([0,1])$, which means that it must be 0 somewhere in the interval $[0,1]$. Where can it touch the x-axis? Not outside the interval I_n, since f_n is positive outside I_n. So, $f^2 + f_n$ is 0 for some $x_n \in I_n$, which in turn means that *both* $f(x_n) = 0$ and $f_n(x_n) = 0$. Thus, for each n, we have an $x_n \in I_n$ for which $f(x_n) = 0$. But $x_n \to p$, since the intervals I_n are nested and their intersection contains only the single point p. Since f is continuous and $x_n \to p$, and $f(x_n) = 0$ for each n, we conclude that $f(p) = 0$. (Recall that a function is continuous at p if and only if $\lim_{x \to p} f(x) = f(p)$.) Therefore, since $f(p) = 0$, $f \in M_p$, and we have shown that $M \subseteq M_p$. But then M is a maximal ideal contained in a maximal ideal M_p, and so $M = M_p$, which is exactly what we wanted to show: *the maximal ideals in $C([0,1])$ turned out after all to be precisely the ones that were so easy to describe from the very beginning.*

We summarize our characterization of the maximal ideals of the ring of continuous functions $C([0,1])$ as follows:

Theorem 7.1. *An ideal M of the ring $C([0,1])$ is maximal if and only if $M = M_p$ for some point $p \in [0,1]$.*

It turns out that this same remarkable result that is true for the closed bounded interval $[0,1]$ is also true for any topological space which is what is called *compact* — that is, a space in which every collection of closed sets has a non-empty intersection provided that every finite subcollection of the closed sets has a non-empty intersection. In particular, it can be shown that a subset of the real line is compact if and only if it is closed and bounded. (This result is called the Heine-Borel theorem.) Thus the set $[0,1]$ is compact, which is the real reason Theorem 7.1 is true.

It is certainly clear that the topological nature of a space X completely determines the ring $C(X)$ of continuous real-valued functions defined on the space X — that is, the ring of all continuous functions $f : X \longrightarrow \mathbf{R}$. Moreover, if X is compact, then the reverse is also true — in other words, the ring $C(X)$ *determines* the topological space X. This says, for example, that all the information about the *topology* of the closed bounded interval $[0,1]$ must somehow be contained within the *algebra* of the ring $C([0,1])$!

The maximal ideals M_p of $C([0,1])$ are what are called *fixed* ideals — ideals such that for some *fixed* point p all the functions in the ideal are 0 at p. Theorem 7.1 can therefore be restated by saying that every maximal ideal M of $C([0,1])$ is a fixed ideal. More generally, it is true that every maximal ideal of $C(X)$ for any compact space X is a fixed ideal.

We end this chapter with an example showing that not all maximal ideals of a ring of continuous functions need be fixed. An ideal that is not fixed (that is, not fixed to a particular p) is called *free*. I might add that it is the presence of free ideals that gives the study of rings of continuous functions its particular appeal.

Example 5

In this example we shall show that $C(\mathbf{R})$, the ring of continuous real-valued functions defined on $\mathbf{R}$, has *free* maximal ideals — that is, not every maximal ideal of $C(\mathbf{R})$ need be of the form M_p for some $p \in \mathbf{R}$. The fixed ideals M_p are still maximal ideals; we are just claiming that not all maximal ideals of $C(\mathbf{R})$ need be of this form.

To this end, let I be the subset of $C(\mathbf{R})$ consisting of those functions that have *bounded support* — that is,

$$I = \big\{ f \in C(\mathbf{R}) \mid f(x) = 0 \text{ for all } x \notin [-n, n] \text{ for some integer } n \big\}.$$

In other words, a function has bounded support if it is 0 *outside* some finite interval. For example, the function $\sin x$ does *not* have bounded support because there are arbitrarily large values of x for which $\sin x$ is nonzero. On the other hand, the function f defined by

$$f(x) = \begin{cases} 3 + 2x - x^2 & \text{for } -1 \le x \le 3, \\ 0 & \text{otherwise} \end{cases}$$

does have bounded support because it is 0 outside the finite interval $[-1, 3]$. You should draw a picture of f, not only to convince yourself that it is continuous, but also because the picture gives you an accurate sense of what functions with bounded support look like.

It is easy to check that the set I is an ideal. Therefore, by Problem 4.1, we know that I is contained in some maximal ideal. But, I cannot be contained in a fixed maximal ideal M_p since, for any given value of p, there are obviously lots of functions in I that are not 0 at p (and hence cannot be in M_p). It is probably quite easy for you simply to draw such a function, but, just to be specific, here is an example that works for any given p:

$$f(x) = \begin{cases} x - p + 1 & \text{for } p - 1 \le x \le p, \\ -x + p + 1 & \text{for } p < x \le p + 1, \\ 0 & \text{otherwise.} \end{cases}$$

Note that $f(p) = 1 \ne 0$, so $f \notin M_p$. Again, you should draw this function to convince yourself that $f \in I$ — that is, that f is continuous and has bounded support. Thus, since I is not contained in a fixed maximal ideal, it must be contained in a free maximal ideal. Hence, free maximal ideals exist.

Problems

7.1   Is the ring $C(\mathbf{R})$ an integral domain?

7.2   Let X be a topological space. (If you don't know what a topological
space is, just think of X as being the real numbers.) For a real-valued
function $f : X \to \mathbf{R}$, define the *zero set* $Z(f)$ of f by

$$Z(f) = \{x \in X \mid f(x) = 0\}.$$

(a) Show that if $f, g \in C(X)$, then

$$Z(f) \cup Z(g) = Z(fg) \text{ and } Z(f) \cap Z(g) = Z(f^2 + g^2).$$

(b) Prove that

$$(f, g) = C(X) \text{ if and only if } Z(f) \cap Z(g) = \emptyset,$$

where (f, g) denotes the ideal of $C(X)$ generated by f and g.

7.3   Let X be a topological space. (Again, if you don't know what a
topological space is, just think of X as being the real numbers.)
Let P and Q be prime ideals of $C(X)$. Prove that

$$PQ = P \cap Q,$$

where PQ is the product of the ideals P and Q (see Problem 3.13).
In particular, then, $P^2 = P$ for any prime ideal P of $C(X)$.

7.4   Let $f \in C([0, 1])$ be the function defined by $f(x) = x - \frac{1}{2}$. Describe the
elements of the principal ideal (f).
        Since $f(\frac{1}{2}) = 0$, the function f is an element of the maximal ideal
$M_{\frac{1}{2}}$. Thus, $(f) \subseteq M_{\frac{1}{2}}$. Show that the ideal (f) is not prime by showing
that $(f)^2 \neq (f)$ and using the result of Problem 7.3. Conclude that
$(f) \subset M_{\frac{1}{2}}$.

7.5   Let I be a proper ideal of $C([0, 1])$. Show that I is a fixed ideal — that is,
show that there is a point $p \in [0, 1]$ such that $f(p) = 0$ for all functions
$f \in I$.

7.6   Give an example of a nested sequence of *open* intervals whose
intersection is a single point.

7.7   The reason that free maximal ideals can exist in $C(\mathbf{R})$ is that $\mathbf{R}$ is not a compact space. A *compact* space is defined to be a space in which every collection of closed sets has a non-empty intersection provided that every finite subcollection of the closed sets has a non-empty intersection. Prove that $\mathbf{R}$ is not compact by finding an infinite collection of closed intervals that has an *empty* intersection and yet any finite number of the intervals have a non-empty intersection.

7.8   In Example 5 we showed that the ideal I of $C(\mathbf{R})$ consisting of all functions of bounded support was contained in a free maximal ideal; call that free maximal ideal M. Show that the ideal I is not prime. Conclude that I is properly contained in M — that is, $I \subset M$.

8

Homomorphisms and Isomorphisms

There have been several situations up to now in which it should have been clear to us that two different rings were in fact really the same ring that had only been presented to us in slightly different forms. For example,

(1) the quotient ring $\mathbf{Z}/(6)$ — whose six elements are the six cosets $0 + (6)$, $1 + (6)$, $2 + (6)$, $3 + (6)$, $4 + (6)$, and $5 + (6)$ — is really the same as the ring of integers $\{0, 1, 2, 3, 4, 5\}$ in which arithmetic is done mod 6;

(2) an integral domain D is really the same as the subring D' of its quotient field F consisting of all equivalence classes of the form $[d/1]$ for $d \in D$;

(3) the total quotient ring $Q(\mathbf{Z} \times \mathbf{Z})$, whose elements look like $\left[\frac{(a,b)}{(c,d)}\right]$, is really the same as the ring $\mathbf{Q} \times \mathbf{Q}$, whose elements look like $\left(\frac{a}{c}, \frac{b}{d}\right)$.

Moreover, there are also situations in which two different rings are really the same ring but just happen to be represented in such radically different forms that it is not at all easy to see that the two rings are in fact really one and the same ring. For example,

(4) the quotient ring $\mathbf{Z}/(12)$ is really the same as the ring $\mathbf{Z}/(3) \times \mathbf{Z}/(4)$;

(5) the familiar ring $\mathbf{C}$ of complex numbers is really the same as the ring $\mathbf{R}[x]/(x^2 + 1)$, which after all is a quotient ring of the polynomial ring $\mathbf{R}[x]$.

What we need, then, is language with which to make these associations precise. In other words, what we need is the concept mathematicians call *isomorphism*, a word that has been coined from two Greek words, *isos* meaning "equal" and *morphe* meaning "form." So, appropriately, *isomorphism* means "equal form." Sometimes, as in the first three cases above, this precision of language will allow us merely to confirm what is already obvious to us, but other times, as in the last two cases, this precision will give us new insights into the nature of the rings we are studying.

More generally, when studying any branch of mathematics, we are interested in those mappings from one object to another object that preserve some or all of the mathematical structure of the objects we are studying. In ring theory, therefore, we pay special attention to those mappings that preserve the algebraic structure of the rings we are studying, that is, to those mappings that respect the operations of addition and multiplication. These mappings are the *homomorphisms*.

This word has also been coined from Greek: *homos* means "same," and so *homomorphism* means "same form." In mathematics we use these Greek roots in such a way that the term isomorphism always carries with it a sense of two mathematical objects being identically the same or equal, whereas, the term homomorphism is somewhat more general in the sense of two objects being very similar in some precise mathematical way, yet not necessarily identical. We begin with a definition of homomorphism.

Definition 8.1. *Let R and S be two rings. A mapping*

$$f : R \longrightarrow S$$

is called a **homomorphism** *if*

> *(1)* $f(x + y) = f(x) + f(y)$;
> *(2)* $f(xy) = f(x) f(y)$;
> *(3)* $f(1) = 1$,

for any elements x and y in R.

If, further, the mapping f is onto the ring S, we say that S is a **homomorphic image** *of R.*

Note in the above definition that the addition and multiplication that appear on the left side of conditions (1) and (2) are taking place in the ring R, whereas the addition and multiplication on the right side is taking place in the ring S. In fact, when we say that a homomorphism *respects* the operations of addition and multiplication, what we mean is that if you add or multiply two elements of R and then map the result to S, you get the same thing as when you first map the two elements to S and then do the addition or multiplication in that ring. Note, similarly, that the 1 on the left side of condition (3) is the identity element of R, whereas the 1 on the right side is the identity element of S.

Since a homomorphism is supposed to preserve algebraic structure, it is not surprising that we require a homomorphism from a ring R to a ring S to map the identity of R to the identity of S. Similarly, we might expect a homomorphism to take the zero element of one ring to the zero element of the other ring, to take additive inverses to additive inverses, and to take multiplicative inverses to multiplicative inverses. By the way, you should recall from Chapter 1 that, by Problem 1.1, additive inverses are unique and so are multiplicative inverses.

So we now verify, therefore, that, in fact, any homomorphism f from a ring R to a ring S has these three desirable properties as well:

(4) $f(0) = 0$;
(5) $f(-x) = -f(x)$ for all $x \in R$;
(6) $f(x^{-1}) = f(x)^{-1}$ for all units $x \in R$.

Proofs

(4) Since f is a homomorphism, $f(0) + f(0) = f(0 + 0) = f(0)$, so $f(0) = 0$;
(5) $f(-x) + f(x) = f(-x + x) = f(0) = 0$, so $f(-x) = -f(x)$;
(6) $f(x^{-1}) f(x) = f(x^{-1}x) = f(1) = 1$, so $f(x^{-1}) = f(x)^{-1}$.

Example 1

Let I be an ideal of a ring R. The natural mapping

$$f : R \longrightarrow R/I$$

defined by

$$f(x) = x + I,$$

which takes an element x of the ring R and sends it to the coset $x + I$ of the ring R/I, is a homomorphism. In fact, since the mapping f is *onto* the ring R/I, R/I is a homomorphic image of R.

The verification of all of this is quite easy. First, we show that f satisfies the three properties required of a homomorphism:

(1) $f(x + y) = (x + y) + I = (x + I) + (y + I) = f(x) + f(y)$;
(2) $f(xy) = xy + I = (x + I)(y + I) = f(x) f(y)$;
(3) $f(1) = 1 + I$, which is the identity of R/I.

We must also show that f is *onto*: but, clearly, f is onto since any element $x + I$ of R/I is the image under f of x.

Surprisingly, it turns out that in a very real sense this simple example of a ring homomorphism is the *only* example possible — in other words, we will soon see that if $f : R \to S$ is an onto homomorphism between two rings, then there is an ideal I of R such that the ring S is really just the same ring as — that is, is isomorphic to — the ring R/I. Finding the ideal I that actually works to show this is quite easy, and this ideal is important enough to deserve a special name:

Definition 8.2. *Let $f : R \to S$ be a homomorphism from a ring R to a ring S. The* **kernel** *of f is the set of elements of R that are mapped to 0*

by f — that is,

$$\ker(f) = \{x \in R \mid f(x) = 0\}.$$

As we indicated above, the kernel of a homomorphism is in fact an ideal. Let $f : R \to S$ be a homomorphism from a ring R to a ring S and let $K = \ker(f)$. We show that K is an ideal of R. Let $a, b \in K$ — that is, $f(a) = 0$ and $f(b) = 0$. Thus $f(a - b) = f(a) - f(b) = 0 - 0 = 0$, and so $a - b \in K$. Let $r \in R$ and $a \in K$; then $f(ra) = f(r)f(a) = f(r) \cdot 0 = 0$, so $ra \in K$. Therefore, $K = \ker(f)$ is an ideal of R.

Returning momentarily to Example 1 above, what is the kernel of the homomorphism f in that example? By definition, $\ker(f)$ is the set of elements of R that get sent by f to the zero element of R/I. So $a \in \ker(f)$ if and only if $f(a) = 0 + I = I$ — that is, if and only if $a + I = I$ or, equivalently, if and only if $a \in I$. Thus, $\ker(f) = I$. In other words, for the natural homomorphism $f : R \to R/I$, the kernel of f is just the ideal I itself.

A special case for the kernel is of particular importance to us. This is when the kernel of a homomorphism consists of just the single element 0. This means that the *only* element that gets sent to 0 is 0. Let's discuss this situation in some detail.

So, suppose $f : R \to S$ is a homomorphism from a ring R to a ring S and that $\ker(f) = \{0\}$. We claim that the homomorphism f is a one-to-one mapping. If $f(x) = f(y)$, for some $x, y \in R$, then $f(x - y) = f(x) - f(y) = 0$, so $x - y \in \ker(f) = \{0\}$; thus, $x - y = 0$, and so $x = y$; we conclude that f is one-to-one. Of course, the converse is obvious: if f is one-to-one, then $\ker(f) = \{0\}$ (since 0 is always sent to 0 by a homomorphism). This result is so fundamental, it is worth repeating our conclusion:

a homomorphism f is one-to-one if and only if $\ker(f) = \{0\}$.

This is a rather remarkable result, for it says that as long as the action of the mapping f is one-to-one at 0, then its action will be one-to-one everywhere!

This brings us to the special case of homomorphism that is so important, that is, when a one-to-one homomorphism is also onto. In this case, we call the homomorphism an *isomorphism*:

Definition 8.3. *An* **isomorphism** *is a homomorphism* $f : R \to S$ *from a ring R to a ring S that is both one-to-one and onto. We say that the rings R and S are* **isomorphic** *and write* $R \cong S$.

The main point, of course, is that if two rings are isomorphic, then we think of them as being the same ring. The isomorphism, or mapping, f between them amounts to nothing more than a *renaming* of the elements — that is, if $f : R \to S$ is an isomorphism, then each element a in R is simply renamed $f(a)$ in the ring S. The element a in R will behave in the ring R exactly in the same way that its renamed counterpart, $f(a)$, will behave in the ring S.

Example 2

Let $f : R \to S$ be an *onto* homomorphism from a ring R to a ring S. Then

$$S \cong R/\ker(f).$$

In other words, *any* homomorphic image S of a ring R is really just a quotient ring of the form R/I for some ideal I.

This important example — really just Example 1 revisited — is known as the **first isomorphism theorem**.

In order to prove this result, we define a mapping

$$\phi : R/\ker(f) \to S$$

by

$$\phi(x + \ker(f)) = f(x)$$

for each $x \in R$. We claim that ϕ is an isomorphism between $R/\ker(f)$ and S. Denote the kernel of f by K. First of all, we should show that ϕ is well defined. We must show that if $x + K = y + K$, then $f(x) = f(y)$. Suppose that $x + K = y + K$. Then $x - y \in K$, and so, $f(x - y) = 0$; that is, $f(x) - f(y) = 0$, since f is a homomorphism. Therefore, $f(x) = f(y)$ and ϕ is well defined.

Next, we show that ϕ is a homomorphism:

(1) $\phi((x + K) + (y + K)) = \phi((x + y) + K) = f(x + y) = f(x) + f(y) = \phi(x + K) + \phi(y + K)$;

(2) $\phi((x + K)(y + K)) = \phi(xy + K) = f(xy) = f(x)f(y) = \phi(x + K)\phi(y + K)$;

(3) $\phi(1 + K) = f(1)$, which is the identity of S since f is a homomorphism from R to S.

Next, we show that ϕ is onto. Let $y \in S$. We must find a *preimage* of y — that is, we must find an element, $x + K$, of R/K such that $\phi(x + K) = y$. But f is onto, so there is an $x \in R$ such that $f(x) = y$. Therefore, $x + K$ is the desired preimage of y, since $\phi(x + K) = f(x) = y$.

Finally, we show that ϕ is one-to-one. It suffices to show that the kernel of ϕ contains the single element K, the zero element of R/K. Suppose that $\phi(x + K) = 0$, which means that $f(x) = 0$, and so, $x \in K$, and $x + K = K$, as desired. Thus, ϕ is an isomorphism, as claimed.

Example 3

Let S be the ring of integers mod 6 — that is, the elements of S are the numbers 0, 1, 2, 3, 4, 5 with the operations of addition and multiplication carried out mod 6. Define a mapping

$$f : \mathbf{Z} \rightarrow S$$

by

$$f(n) = \text{the remainder when } n \text{ is divided by 6.}$$

It is easy to check that f is an onto homomorphism and that $\ker(f) = (6)$. Therefore, the ring $\mathbf{Z}/(6)$ is isomorphic to the ring of integers mod 6.

We end this chapter with two examples to illustrate that these concepts of homomorphism and isomorphism are more than just a convenience of language. They give us a powerful new way of looking at things.

Example 4

For some fixed $p \in [0, 1]$, let

$$\phi : C([0, 1]) \rightarrow \mathbf{R}$$

be the mapping defined by

$$\phi(f) = f(p).$$

The mapping ϕ from the ring of continuous functions to the real numbers is called the *evaluation map* at p. Since ϕ is an onto homomorphism (this is easy to check), $\mathbf{R}$ is isomorphic to $C([0, 1])/\ker(\phi)$, which means that $\ker(\phi)$ must be a maximal ideal since the ring $\mathbf{R}$ itself is a field.

But

$$\ker(\phi) = \{f \in C([0, 1]) \mid \phi(f) = 0\}$$

$$= \{f \in C([0, 1]) \mid f(p) = 0\}$$

$$= M_p.$$

Thus, we could conclude that M_p is a maximal ideal, an important fact concerning the ring $C([0, 1])$ that we verified in an entirely different way in the last chapter.

Example 5

Let

$$\phi : \mathbf{R}[x] \to \mathbf{C}$$

be the mapping defined by

$$\phi(f) = f(i).$$

Since ϕ is an onto homomorphism from the ring of polynomials with real coefficients to the complex numbers, we see that the field $\mathbf{C}$ is isomorphic to $\mathbf{R}[x]/\ker(\phi)$. Note that the polynomial $x^2 + 1$ is in the kernel of ϕ since $i^2 + 1 = 0$. In fact, it turns out that $\ker(\phi) = (x^2 + 1)$. Therefore,

$$\mathbf{C} \cong \mathbf{R}[x]/(x^2 + 1),$$

which gives us an entirely new way of looking at the complex numbers — namely, as a quotient ring of the polynomial ring $\mathbf{R}[x]$.

Problems

8.1   Let $f : \mathbf{Z} \to \mathbf{Z} \times \mathbf{Z}$ be the mapping defined by $f(n) = (n, 0)$. Is f a homomorphism?

8.2   Let D be an integral domain and let K be its quotient field. Find a subring D' of K such that $D \cong D'$. (This, for example, is our justification for thinking of $\mathbf{Z}$ being contained in $\mathbf{Q}$.)

8.3   Show that $Q(\mathbf{Z} \times \mathbf{Z}) \cong \mathbf{Q} \times \mathbf{Q}$.

8.4   Show that $\mathbf{Z}/(12) \cong \mathbf{Z}/(3) \times \mathbf{Z}/(4)$.

8.5   Let $\mathbf{Z}[\sqrt{2}]$ be the ring of all real numbers of the form $a + b\sqrt{2}$, where a and b are integers. (It is easy to show that $\mathbf{Z}[\sqrt{2}]$ is a subring of $\mathbf{R}$.)
    Let

$$f : \mathbf{Z}[\sqrt{2}] \to \mathbf{Z}[\sqrt{2}]$$

be the mapping defined by

$$f(a + b\sqrt{2}) = a - b\sqrt{2}.$$

Show that f is an isomorphism.

8.6   Let $f : R \to S$ be a homomorphism from a ring R to a ring S. We define the **image** of f to be the set

$$f(R) = \{f(x) \mid x \in R\}.$$

We sometimes use the notation im(f) for this set.
    Show that the image $f(R)$ is a subring of S.

8.7   Let

$$\phi : \mathbf{Z}[x] \to \mathbf{C}$$

be the mapping defined by

$$\phi(f) = f(i).$$

What familiar subring of $\mathbf{C}$ is the image $\phi(\mathbf{Z}[x])$ in this case?

8.8   Let $f : R \to S$ be an isomorphism from a ring R to a ring S. Show that the inverse mapping $f^{-1} : S \to R$ is also an isomorphism. (The inverse mapping is defined as follows: $f^{-1}(s) = r$ provided that $f(r) = s$.)

8.9   Homomorphisms and isomorphisms are supposed to preserve algebraic structure. However, not all algebraic properties of a ring need be preserved under homomorphism. Find an example of an integral domain that has a homomorphic image that is not an integral domain. Are *all* algebraic properties of a ring preserved under isomorphism?

8.10   Let R be a nonzero ring. Prove that R is a field if and only if every homomorphism from R onto any nonzero ring is an isomorphism.

8.11   Let $f : R \to S$ be an onto homomorphism from a ring R to a ring S. Prove that there is a one-to-one, order-preserving correspondence between the *ideals of S* and the *ideals of R that contain* $\ker(f)$.

8.12   **The second isomorphism theorem**. Let I and J be ideals of a ring R such that $I \subseteq J$. First, explain what we mean by the set J/I, then show that J/I is an ideal of R/I, and finally use the first isomorphism theorem as described in Example 2 on page 84 to prove that

$$R/I \big/ J/I \cong R/J.$$

8.13   **The third isomorphism theorem**. Let R be a subring of a ring S, and let J be an ideal of S. First show that $J \cap R$ is an ideal of R, then show that $J + R$ is a subring of S, and finally use the first isomorphism theorem as described in Example 2 on page 84 to prove that

$$R/J \cap R \cong J + R/J.$$

9

Unique Factorization

Unique Factorization Domains

In the next two chapters we explore several of the possible general-izations of the ring of integers $\mathbf{Z}$. One generalization of $\mathbf{Z}$ that we have already seen is that of an integral domain, a generalization which preserves the property that any two nonzero elements in the ring must have a nonzero product. In the case of this particular generalization, of course, the name itself serves to remind us of this notion of general-ization: *integral* domain. The integers also have other properties which may, or may not, in general be shared by other integral domains — a fact that several great mathematicians were quite slow to realize, and which has given this topic a rich and fruitful history.

Certainly, one of the fundamental notions involving the integers is that of divisibility. As was suggested by our discussion of prime ideals, all the information about divisibility in a ring is contained within the ideal structure of the ring. This can be made explicit by showing that the divisibility of two elements is reflected in the containment of the principal ideals generated by the two elements.

Let R be a ring and let $x, y \in R$. Recall that we write $x|y$ and say "x divides y" if $y = rx$ for some element $r \in R$. Thus we have

$$x|y \text{ if and only if } (y) \subseteq (x).$$

In particular,

$$x|y \text{ and } y|x \text{ if and only if } (x) = (y).$$

For example, in the ring of integers $\mathbf{Z}$, if two integers x and y divide one another, then $y = x$ or $y = -x$; that is, if two ideals of $\mathbf{Z}$ are such that $(x) = (y)$, then $y = x$ or $y = -x$. However, in an arbitrary ring there can be more than two elements that divide one another. For an extreme example, in a field every pair of nonzero elements divide one another. This leads us to the following definition.

Definition 9.1. *Let x and y be elements of a ring R. We say that x and y are* **associates** *if $x = uy$ for some unit $u \in R$.*

Note that if x and y are associates in a ring R, then $(x) = (y)$: assume that $x = uy$ for some unit $u \in R$, thus $y|x$; but u is a unit, so $y = u^{-1}x$; thus, $x|y$; therefore, $(x) = (y)$.

The converse, quite surprisingly, is not true in general. A simple example can be given in the ring $\mathbf{Z}/(6)$:

$$(2 + (6)) = \{0 + (6), \ 2 + (6), \ 4 + (6)\} = (4 + (6)),$$

but $2 + (6)$ and $4 + (6)$ are not associates in $\mathbf{Z}/(6)$.

However, in an integral domain, this converse *is* true, and so, for two elements x and y in an *integral domain D*, we have that

$$x \text{ and } y \text{ are associates if and only if } (x) = (y).$$

In order to prove this, we need only prove the converse half, so we assume that $(x) = (y)$: then $x = ry$ and $y = sx$ for some $r, s \in D$; thus, $x = ry = rsx$; in other words, $x(1 - rs) = 0$; since D is an integral domain, either $x = 0$ or $1 - rs = 0$; if $x = 0$, then $y = 0$ and x and y are associates; if $1 - rs = 0$, then r is a unit, and x and y are associates.

Certainly one of the most useful and interesting properties of the integers is the property known to us as the *fundamental theorem of arithmetic*, which tells us that every integer greater than 1 can be uniquely factored into a product of prime numbers. For example, 1729 can be written as $7 \cdot 13 \cdot 19$. The whole point about *unique* factorization, of course, is that this is the *only* way to factor 1729 into a product of primes. Naturally, the crucial property of prime numbers we are using in this context is that they can't be factored any further than they already are — they are, quite simply, *irreducible*.

Definition 9.2. *A nonzero, non-unit element x of a ring R is said to be* **irreducible** *if its only divisors are units and associates of x — that is, if $x = rs$ for some $r, s \in R$, then one of r and s is a unit.*

So the most familiar and basic property of primes numbers in the ring of integers — namely, their irreducibility — is the one we just recorded. But another crucial property of prime numbers is the one we used in Chapter 3 to motivate the definition of prime ideal — namely, the property that an integer p is prime if and only if whenever $p|ab$ for two integers a and b, then $p|a$ or $p|b$. We now use this property of prime integers to give a different generalization of what it means for an element in a ring to be *prime*.

Definition 9.3. *A nonzero element x of a ring R is said to be* **prime** *if the principal ideal (x) is prime, that is, if $ab \in (x)$ for some $a, b \in R$, then $a \in (x)$ or $b \in (x)$.*

In the ring of integers **Z**, these two concepts coincide, that is, an integer is irreducible *if and only if* it is prime. This is not true in general. In fact, as we show in the next two examples, there are rings in which there are prime elements that are not irreducible, and there are also rings in which there are irreducible elements that are not prime.

Example 1

In the ring $\mathbf{Z}/(6)$, the element $3 + (6)$ is prime but not irreducible. It is not irreducible, since we can write $3 + (6) = (3 + (6))(3 + (6))$, and this is a nontrivial factorization since $3 + (6)$ is not a unit. However, we can still show that $3 + (6)$ is a prime element. In order to see this, consider the ideal $(3 + (6)) = \{0 + (6), 3 + (6)\}$. If $(a + (6))(b + (6)) \in (3 + (6))$, then $ab + (6) \in (3 + (6))$, so $3|ab$, but then $3|a$ or $3|b$, which means that one of the elements $a + (6)$ or $b + (6)$ is in $(3 + (6))$. Thus $3 + (6)$ is prime.

It might be worth pausing to note that things have gotten quite bizarre in this example. We are used to the idea in the integers that if we begin with an integer n, either it is prime already or else it can be factored nontrivially into $n = ab$. Then either the integers a and b are prime or they also can be factored, and so on. Continuing this process eventually must result in the complete factorization of the original integer n into a product of primes. But, look what happens in Example 1 if we try to repeat this simple factorization process on $3 + (6)$. Since $3 + (6) = (3 + (6))(3 + (6))$, we see that one could continue this process of factoring forever!

Example 2

Let $\mathbf{Z}[\sqrt{-5}] = \{a + b\sqrt{-5} \mid a, b \in \mathbf{Z}\}$. Then $\mathbf{Z}[\sqrt{-5}]$ is a subring of the complex numbers **C**. The element $2 + \sqrt{-5}$ is an irreducible element (we shall see how to verify this in Problem 9.3), but it is not a prime element since $2 + \sqrt{-5} \mid 3 \cdot 3$, and yet $2 + \sqrt{-5} \nmid 3$. We see that $2 + \sqrt{-5} \mid 9$, since $(2 - \sqrt{-5})(2 + \sqrt{-5}) = 9$. On the other hand, if $2 + \sqrt{-5} \mid 3$, then $(a + b\sqrt{-5})(2 + \sqrt{-5}) = 3$ for some $a, b \in \mathbf{Z}$. Then $2a - 5b = 3$ and $a + 2b = 0$. Solving these equations yields $a = \frac{2}{3}$ and $b = -\frac{1}{3}$, contradicting the fact that a and b are supposed to be integers.

An obvious question now arises: are there any restrictions we can place on a ring so that — as was the case for the ring of integers — the two concepts of prime elements and irreducible elements will once again coincide? In Example 1, the ring $\mathbf{Z}/(6)$ has zero-divisors, which, as you may have guessed, was what allowed us to find a prime element in that particular ring that is not irreducible.

However, if we insist that a ring be an *integral domain*, it will once again be true that

every prime element is irreducible.

In order to prove this, let x be a prime elment of an integral domain D. Suppose that $x = rs$ for some $r, s \in D$. Since x is prime, either $r \in (x)$ or $s \in (x)$. We may as well assume that $r \in (x)$. So $r = tx$ for some $t \in D$. Thus, $x = tsx$, and so $x(1 - ts) = 0$. Since D is a domain and $x \neq 0$, we conclude that $1 - ts = 0$, and so s is a unit. Therefore, x is irreducible.

In the other direction, if we further require a ring to have — as do the integers — the property of *unique factorization*, then, as we shall see in Problem 9.6, it is also true in such a ring that

every irreducible element is prime.

We give a special name to those integral domains which share with the integers this vitally important property of unique factorization.

Definition 9.4. *An integral domain D is called a* **unique factorization domain** *if every nonzero, non-unit element of D has a unique factorization into a product of irreducible elements of D.*

This overly long but highly descriptive terminology is often somewhat clumsily shortened by referring to such a domain D as a UFD. Of course, we still need to clarify what we mean by *unique factorization*. By saying that a factorization is *unique*, we mean *unique up to order and multiplication by units*. For example, we think of the factorizations

$$12 = 2^2 \times 3 = 3 \times 2 \times 2 = -3 \times -2 \times 2 = 2 \times -1 \times 3 \times -2$$

in the ring of integer $\mathbf{Z}$ as all being essentially the same factorization of the integer 12.

More generally, we think of a factorization

$$x = p_1 p_2 p_3$$

in a ring as being the same as the factorization

$$x = p_2(up_1)(u^{-1}p_3),$$

where u is a unit and the p_i are irreducible elements. In other words, we are not concerned with the order of the factors, nor do we care whether a particular factor is represented by an irreducible element p_i or by one of its associates up_i.

More precisely, we call two factorizations

$$x = p_1 p_2 \cdots p_m = q_1 q_2 \cdots q_n$$

equivalent if there is a one-to-one and onto function

$$f : \{p_i\} \longrightarrow \{q_i\}$$

such that p_i and $f(p_i)$ are associates for each i.

We have already seen an example of a ring in which unique factorization *fails*:

Example 3

The ring $\mathbf{Z}[\sqrt{-5}]$ from Example 2 on page 91 is not a unique factorization domain since, for example, the element 9 has two distinct factorizations,

$$9 = 3 \times 3 = (2 + \sqrt{-5}) \times (2 - \sqrt{-5}).$$

In order to show that these two factorizations are not equivalent, we must show that 3 is not an associate of $2 + \sqrt{-5}$ or $2 - \sqrt{-5}$. Assume that $3 = (a + b\sqrt{-5})(2 + \sqrt{-5})$, where $a + b\sqrt{-5}$ is a unit in $\mathbf{Z}[\sqrt{-5}]$, and $a, b \in \mathbf{Z}$ — that is, assume that 3 and $2 + \sqrt{-5}$ *are* associates. Then, $2a - 5b = 3$ and $a + 2b = 0$. Solving these equations yields $a = \frac{2}{3}$ and $b = -\frac{1}{3}$, contradicting the fact that a and b are supposed to be integers. A similar contradiction arises if we assume that 3 and $2 - \sqrt{-5}$ are associates. For the final details of showing in this example that 3, $2 + \sqrt{-5}$, and $2 - \sqrt{-5}$ are actually irreducible elements, see Problem 9.3.

Another example of a domain that is not a unique factorization domain is given in Problem 9.11.

Fermat's Last Theorem

This failure of unique factorization is an astounding phenomenon. The slow realization among mathematicians in the nineteenth century that such pathology can actually occur in mathematical systems is closely tied to the story of *Fermat's Last Theorem*.

Pierre de Fermat was one of the greatest mathematicians of the seventeenth century, in spite of the fact that he was by occupation a jurist and rarely published any mathematics. Yet he laid much of the preliminary groundwork for calculus with his work on tangents and on maxima and minima problems. There is a Fermat's *principle of least time*, so called because he showed that the principle that "light travels from point *A* to point *B* by the path that minimizes time" implies the familiar law of refraction. But, above all, he is remembered for his marvelous work in number theory.

Fermat himself was perhaps proudest of his discovery of the *method of infinite descent*. Suppose that we wish to show that no positive integer has a certain property. We can begin by assuming that some positive integer has the given property. Then, we proceed to show how to find a smaller positive integer that also has the given property. Since this process could now be repeated, each time yielding smaller and yet smaller positive integers having the given property, this gives us an *infinite descent* of positive integers, each of which has the given property, and this is clearly impossible. This contradiction shows that no positive integer could have the given property.

Fermat was famous in his day because of his correspondence with other mathematicians, and yet he hardly ever published any of his discoveries. However, among Fermat's work — published only after his death in 1665 — were the notes he had written in the margins of his own copy of Bachet's translation of Diophantus's *Arithmetica*. It was Bachet's Latin translation of Diophantus — published in 1621 — that was the fertile source of much of Fermat's work in number theory. In particular, Diophantus's book contained the following problem: given a number which is a square, write it as the sum of two other squares. In the margin next to this problem, Fermat had written:

> On the other hand, it is impossible for a cube to be written as a sum of two cubes or a fourth power to be written as a sum of two fourth powers or, in general, for any number which is a power greater than the second to be written as a sum of two like powers. I have a truly marvelous demonstration of this proposition which this margin is too narrow to contain.

This enigmatic last sentence made Fermat's Last Theorem one of the most famous and most tantalizing problems in mathematics, one that remained unsolved for over 350 years! The original problem of Diophantus was the relatively easy problem of characterizing what are called *Pythagorean triples* — that is, finding all positive integer or, equivalently, rational solutions to the equation

$$x^2 + y^2 = z^2,$$

such as $x = 3$, $y = 4$, $z = 5$ or $x = 5$, $y = 12$, $z = 13$. What we now call Fermat's Last Theorem is really his extremely bold *conjecture* that, for $n > 2$, the equation

$$x^n + y^n = z^n$$

has no positive integer or rational solutions.

In fact, we clearly need be concerned only with integer solutions to this equation. For suppose that three rational numbers $\frac{a}{b}$, $\frac{c}{d}$, and $\frac{e}{f}$ provide a solution to this equation, where $a, b, c, d, e, f \in \mathbf{Z}$. Then,

$$\left(\frac{a}{b}\right)^n + \left(\frac{c}{d}\right)^n = \left(\frac{e}{f}\right)^n.$$

If we multiply this equation through by $(bdf)^n$, we get

$$(adf)^n + (cbf)^n = (ebd)^n,$$

and we have produced an integer solution to the original equation. So, if the equation $x^n + y^n = z^n$ has no integer solutions for any $n > 2$, then it can have no rational solutions either.

In 1994, more than 350 years after Fermat wrote his famous marginal note in his copy of *Arithmetica*, Fermat's Last Theorem was finally proven. Although Andrew Wiles actually announced his proof in the summer of 1993 at a series of lectures in Cambridge, England, a serious flaw in the proof was discovered in September, and it took another year, and help from former student Richard Taylor, for Wiles to patch up the final proof.

The proof by Wiles was in fact but the final piece in a very complicated story that began in the 1950s with a conjecture that connects topology and number theory called the Taniyama-Shimura conjecture. This conjecture was first connected to Fermat's Last Theorem by Gerhard Frey in 1984. The really drammatic breakthrough came in 1986 when Ken Ribet proved that the Taniyama-Shimura conjecture implies Fermat's Last Theorem. Thus Wiles's final assault on Fermat's Last Theorem was his ultimately successful attack on a special case of the Taniyama-Shimura conjecture.

Progress through the centuries on Fermat's Last Theorem indeed makes a fascinating tale. Fermat himself was able to do the special case $n = 4$. It was another century before Leonard Euler did the case $n = 3$. In about 1825 Sophie Germain was the first to prove it, not for just a single value of n such as $n = 3$ or $n = 4$, but for an entire class of numbers, the numbers we now call *Germain primes* — a prime p is a *Germain prime* if both p and $2p + 1$ are prime. Only in 1983 was Gerd

Faltings able to prove that the theorem must be true for infinitely many prime values of n.

Let us now go back and take a look at Euler's "proof" of the case $n = 3$, which is interesting because the gap in his argument has to do with unique factorization, our main topic in this chapter. Euler used an infinite descent argument and arrived at a point where he observed that if two integers p and q are of the form $p = a^3 - 9ab^2$ and $q = 3a^2b - 3b^3$, then $p^2 + 3q^2 = (a^2 + 3b^2)^3$ — you can check Euler's algebra if you want. In other words, $p^2 + 3q^2$ is a cube. But then, as part of Euler's proof, it becomes necessary to know that this is the *only* way in which $p^2 + 3q^2$ can be expressed as a cube.

Euler gave a brilliant — albeit fallacious — argument for this by using numbers of the form $a + b\sqrt{-3}$, where a and b are integers. Thus, $p^2 + 3q^2 = (p + q\sqrt{-3})(p - q\sqrt{-3})$, and so, if $p^2 + 3q^2$ is a cube (said Euler), then each of its factors must also be a cube. Thus, suppose that $p + q\sqrt{-3} = (a + b\sqrt{-3})^3$. Then, easily, $p - q\sqrt{-3} = (a - b\sqrt{-3})^3$, and so we conclude that $p^2 + 3q^2 = (a + b\sqrt{-3})^3(a - b\sqrt{-3})^3 = ((a + b\sqrt{-3})(a - b\sqrt{-3}))^3 = (a^2 + 3b^2)^3$, and this is the only way $p^2 + 3q^2$ can be expressed as a cube, just as Euler claimed.

As part of his argument, then, Euler argued that in order for a number $x^2 + cy^2$ to be a cube it is necessary that both of its complex factors, $x + y\sqrt{-c}$ and $x - y\sqrt{-c}$, must also be cubes, since they are relatively prime in that x and y have no common divisors. He offered no proof, however. In fact, this statement happens to be true for the particular ring that he was using. It is possible that Euler had a more pedestrian way of showing that if $x^2 + 3y^2$ is a cube, then $x = a^3 - 9ab^2$ and $y = 3a^2b - 3b^3$, and that only later did he give the elegant but fallacious argument described above using complex numbers.

Euler was undoubtedly one of the greatest and most prolific mathematicians of all time. So, if Euler expected other number systems to behave as well as the integers do, it is not surprising that the rest of us should have similar expectations. Of course, the fact that other number systems do not always behave in the way we expect is precisely what makes ring theory so much fun.

The case $n = 5$ of Fermat's Last Theorem was settled in 1825 by Lejeune Dirichlet, who was twenty years old at the time, and by Adrien-Marie Legendre, who was seventy. Dirichlet also settled the case $n = 14$, but failed with the case $n = 7$. This case was done by Gabriel Lamé in 1839. Then, in the early spring of 1847, Lamé dramatically announced to the Paris Academy that he had proved Fermat's Last Theorem. Of course, we know that he had done no such thing, but the story of what happened next is interesting, nonetheless.

The key, said Lamé, was to use the idea, hardly new, that if r is a complex nth root of unity — that is, $r^n = 1$, where $r \in \mathbf{C}$ — then

$$x^n + y^n = (x + y)(x + ry)(x + r^2 y) \cdots (x + r^{n-1} y).$$

You might want to see how this works by trying, for example, the case $n = 3$ and letting r be a cube root of unity, that is, $r^3 = 1$. The key is to note that, if $r \neq 1$, then $1 + r + r^2 = 0$, since $(r - 1)(r^2 + r + 1) = r^3 - 1 = 0$.

Therefore, Lamé's idea was to say that if $x^n + y^n$ was an nth power, z^n, then showing each pair of factors to be relatively prime would mean that each factor would have to be an nth power. This would set up an infinite descent.

Joseph Liouville rose to his feet to point out the flaw in the argument: must each factor be an nth power if the factors are relatively prime and their product is an nth power? That this is true for the integers is because of the possibility of factorization into primes. Not only had Liouville seen the flaw, but he understood that unique factorization was at the core of it.

Thus began a flurry of activity that lasted for months. Augustin-Louis Cauchy supported the approach of Lamé, who himself saw no insurmountable obstacle, and Cauchy produced his own verification on March 22. On that same day both he and Lamé deposited "secret packets" with the Academy as a hedge against any disputes about priority that might arise. Then they each published still more notices. Finally, on May 24, and none too soon it seems, word arrived from the central character in this drama.

A letter from Ernst Eduard Kummer was read to the Academy by Liouville: *Unique factorization fails!* Kummer in fact had published this three years earlier. He then went on to propose using his *ideal complex numbers*, about which he had written the previous year. It was these *ideal* numbers that were later to motivate the naming of Dedekind's ideals. Kummer's main interest was elsewhere, and for him Fermat's Last Theorem was just a passing curiosity. Nevertheless, by April 11, Kummer had been able to prove the theorem for all of what are called the *regular* primes, which, for example, took care of all primes less than 100 except for 37, 59, and 67.

Problems

9.1   Let R be a ring. Let $\cong$ be a relation defined by

$$x \cong y \text{ if and only if } x \text{ and } y \text{ are associates}$$

for $x, y \in R$. Show that $\cong$ is an equivalence relation on R.

9.2   Show that $\mathbf{Z}[\sqrt{-5}] = \{a + b\sqrt{-5} \mid a, b \in \mathbf{Z}\}$ is a subring of $\mathbf{C}$. (Hint: you can do this directly *or* you can express $\mathbf{Z}[\sqrt{-5}]$ as a homomorphic image of $\mathbf{Z}[x]$.)

9.3   In this problem we complete the details of our discussion — begun in Examples 2 and 3 — of the ring $\mathbf{Z}[\sqrt{-5}]$.
       Let $a + bi \in \mathbf{C}$. We define the *norm* of $a + bi$ to be

$$N(a + bi) = a^2 + b^2.$$

First, show that $N(z_1 z_2) = N(z_1)N(z_2)$, for any $z_1, z_2 \in \mathbf{C}$ (see Problem 5.7).
       Then, show that

$$u \text{ is a unit in the ring } \mathbf{Z}[\sqrt{-5}] \text{ if and only if } N(u) = 1.$$

Finally, show that 3, $2 + \sqrt{-5}$, and $2 - \sqrt{-5}$ are irreducible elements of $\mathbf{Z}[\sqrt{-5}]$.

9.4   We will see in the next chapter that the ring $\mathbf{Z}[i]$ of Gaussian integers is a unique factorization domain. With this in mind, explain how it is still possible for 5 to have the following two factorizations:

$$5 = (2 + i)(2 - i) = (1 + 2i)(1 - 2i).$$

9.5   Show that 2 is *not* an irreducible element of $\mathbf{Z}[i]$.

9.6   Let x be an element of a unique factorization domain D. Prove that

$$x \text{ is prime if and only if } x \text{ is irreducible.}$$

9.7   Show that in order to prove Fermat's Last Theorem it is sufficient to prove it for the case $n = 4$ and for the case of all odd primes. This, of course, is why all the attention in Fermat's Last Theorem has been on primes.

9.8   In Euler's "proof" of the case $n = 3$ of Fermat's Last Theorem, he set
        $p + q\sqrt{-3} = (a + b\sqrt{-3})^3$. Verify that $p - q\sqrt{-3} = (a - b\sqrt{-3})^3$, and
        that $p = a^3 - 9ab^2$ and $q = 3a^2b - 3b^2$.

9.9   Let r be a complex nth root of unity. Verify the formula

$$x^n + y^n = (x + y)(x + ry)(x + r^2y) \cdots (x + r^{n-1}y).$$

        (Hint: let $r = e^{2\pi i/n}$, and factor the polynomial $x^n - 1$ into linear
        terms.)

9.10   Let x and y be two elements of a ring R. A **greatest common divisor**
        of x and y, which we write as $\gcd(x, y)$, is an element d of R such that d
        is a common divisor of x and y and, further, any other common divisor
        of x and y must also divide d. Show that if D is a unique factorization
        domain, then any two elements of D have a greatest common divisor.

9.11   Let R be the ring of all polynomials with integer coefficients that have
        no x term; that is, all polynomials of the form

$$a_0 + a_2x^2 + a_3x^3 + \cdots + a_nx^n.$$

        This particular polynomial ring is usually denoted by $R = \mathbf{Z}[x^2, x^3]$.
            Find $\gcd(x^2, x^3)$ and $\gcd(x^5, x^6)$ in $R = \mathbf{Z}[x^2, x^3]$ (see Problem 9.10).
        Also, show that $R = \mathbf{Z}[x^2, x^3]$ is not a unique factorization domain by
        finding two distinct factorizations for the polynomial x^6.

10

Euclidean Domains and Principal Ideal Domains

Principal Ideal Domains

In this chapter we continue with our process of identifying fundamental properties which — like unique factorization — integral domains can share with the ring of integers. One such property is a property of the ring $\mathbf{Z}$ that we first discussed back in Chapter 2 in Example 7, namely, the property that every ideal of $\mathbf{Z}$ is a principal ideal. This remarkable property of the ring $\mathbf{Z}$ suggests the following definition.

Definition 10.1. *An integral domain D is called a* **principal ideal domain** *— or PID, for short — if every ideal of D is principal.*

Example 1

For the record, we repeat the conclusion of Example 7 in Chapter 2, namely,

$$\mathbf{Z} \text{ is a principal ideal domain.}$$

In particular, if I is a nonzero ideal of $\mathbf{Z}$, then $I = (n)$, where n is the least positive integer in I.

Example 2

We will see — as a result of Problem 10.1 — that if K is a field, then the polynomial ring

$$K[x] \text{ is a principal ideal domain.}$$

In particular, if I is a nonzero ideal of $K[x]$, then $I = (f)$, where f is a polynomial of least positive degree in I.

So, $\mathbf{Q}[x]$, $\mathbf{R}[x]$, and $\mathbf{C}[x]$ are all principal ideal domains, as is $\mathbf{Z}/(p)[x]$ for any prime p.

Example 3

The polynomial ring $\mathbf{Z}[x, y]$ is *not* a principal ideal domain since the ideal (x, y) generated by the polynomial x and the polynomial y is not a principal ideal (see Problem 2.18).

So, now, let's find out what, if any, interesting features of the ring of integers have been preserved by our generalization to principal ideal domains. For example, we know that in an arbitrary ring any maximal ideal is also a prime ideal. We also know that in the integers the converse is true as well; that is, any nonzero prime ideal is also a maximal ideal. This happens to be one feature of the integers that is shared by any principal ideal domain, as we now show.

Let D be a principal ideal domain and let (p) be a nonzero prime ideal of D, where $p \in D$. We show that (p) is maximal. Since (p) is a prime ideal, p is a prime element, but then (as we saw in the last chapter) p is irreducible. Now, suppose (p) is properly contained in an ideal (r) of D. Then $r|p$, but p is irreducible and r and p are not associates (since $(r) \neq (p)$), so r must be a unit. Thus, $(r) = D$, and (p) is a maximal ideal, as desired. Therefore, just as is the case in the ring of integers, in any *principal ideal domain*,

every nonzero prime ideal is a maximal ideal.

For a similar example, now looking at elements, we know that in an arbitrary integral domain every prime element is irreducible. We also know that in the integers these two concepts coincide. Not too surprisingly, this feature of the integers is also shared by any principal ideal domain, as we now show.

Let D be a principal ideal domain and let p be an irreducible element of D. We show that p is a prime element. In order to show that p is prime we must show that if $p|rs$ for two elements r and s in D, then $p|r$ or $p|s$. Write $pt = rs$ for some $t \in D$. Since D is a principal ideal domain, $(p, r) = (a)$ for some $a \in D$. So $p = ba$ for some $b \in D$. But, p is irreducible, so either b or a is a unit. We consider each case in turn.

If b is a unit, then $a = b^{-1}p$. Thus $(p, r) = (a) \subseteq (p)$, which means that $r \in (p)$. It follows that $p|r$.

If, on the other hand, a is a unit, then $(a) = D$, so $(p, r) = D$ and $1 = cp + dr$ for some $c, d \in D$. Thus $s = cps + drs = cps + dpt$, and so $p|s$. Therefore, p is prime, since $p|r$ or $p|s$, as desired. We conclude that, just as is the case in the ring of integers, in any *principal ideal domain*

every irreducible element is a prime element.

What about factorization in a principal ideal domain? Since factorization is such a fundamentally important property of the integers, we certainly have every right to hope that principal ideal domains have some nice factorization properties as well. Let us recall again how factorization works in the integers. You simply take an integer and start factoring. If it is irreducible — that is, prime — you stop. Otherwise, you write it as a product of two "smaller" integers. You just keep going until all the factors are irreducible. It is the fact that the factors get "smaller" at each stage that guarantees we will eventually be able to stop. Contrast this with the situation in the ring $\mathbf{Z}/(6)$, where $3 + (6)$ factors nontrivially into $(3 + (6)) \cdot (3 + (6))$. We could now factor each of these terms, thus beginning a process of factoring that we have no hope of ever finishing.

So, filled with optimism, let's see what happens if we try the same thing in a principal ideal domain. Let D be a principal ideal domain and let x_0 be a nonzero, non-unit element of D. If x_0 is not irreducible, then it has a proper divisor — that is, a divisor that is neither a unit nor an associate. Call that proper divisor x_1. If x_1 is not irreducible, then it also has a proper divisor x_2. This process must eventually stop somewhere if we are to have any hope of expressing x_0 as a product of irreducible elements. Another way to look at this is that this string of divisors we are getting gives us what is called an *ascending* chain of ideals of D:

$$(x_0) \subset (x_1) \subset (x_2) \subset \cdots .$$

We just hope this chain will end. Let's see whether it will.

Now, the union $\bigcup_{i \geq 0}(x_i)$ of this ascending chain of ideals is itself an ideal of D such that

$$(x_0) \subset (x_1) \subset (x_2) \subset \cdots \subseteq \bigcup_{i \geq 0}(x_i).$$

However, since we are in a principal ideal domain D, $\bigcup_{i \geq 0}(x_i) = (x)$, for some element $x \in D$ — that is, $\bigcup_{i \geq 0}(x_i)$ is a principal ideal. So

$$(x_0) \subset (x_1) \subset (x_2) \subset \cdots \subseteq (x).$$

But $x \in \bigcup_{i \geq 0}(x_i)$, and so $x \in (x_k)$ for some $k \geq 0$; therefore, $(x) = (x_k)$, since $(x_k) \subseteq (x)$ and $x \in (x_k)$. Thus, this ascending chain of ideals actually terminates at (x_k), and x_k is an irreducible factor of x_0, just as we hoped.

We conclude that x_0 is itself irreducible, or has an irreducible factor, which we call d_1 (that is, the irreducible factor x_k found above). Repeat the argument, if necessary, for $\frac{x_0}{d_1}$ to get another irreducible factor d_2. Continuing in this way we get yet another ascending chain of ideals of D:

$$(x_0) \subset \left(\frac{x_0}{d_1}\right) \subset \left(\frac{x_0}{d_1 d_2}\right) \subset \cdots ,$$

a chain that also must terminate — say, after k steps — for the same reason as above. So, $\frac{x_0}{d_1 d_2 \cdots d_k}$ is irreducible, and we have succeeded in expressing x_0 as a product of irreducible elements, namely,

$$x_0 = d_1 d_2 \cdots d_k \left(\frac{x_0}{d_1 d_2 \cdots d_k} \right).$$

In summary, then, we have shown that in any principal ideal domain, just as is the case in the ring of integers, any nonzero, non-unit element can be factored as a product of irreducible elements. Of course, the big remaining question is whether this factorization must be unique. Remarkably, the answer is *yes*.

Theorem 10.1. *Every principal ideal domain is a unique factorization domain.*

Proof
We have already shown above that any nonzero, nonunit element in a principal ideal domain can be factored as a product of irreducible elements. We complete the proof by showing that, in a principal ideal domain, factorization of an element into a product of irreducible elements is unique.

Suppose that we have two factorizations of an element x into a product of irreducible elements

$$x = p_1 p_2 p_3 \cdots p_m = q_1 q_2 q_3 \cdots q_n,$$

with, say, $m \leq n$. Since every irreducible element in a principal ideal domain is prime, the irreducible elements p_i and q_i are prime, for all i. Since $p_1 | q_1 q_2 \cdots q_n$, and p_1 is prime, p_1 divides one of the q_i. Without loss of generality, we may assume that $p_1 | q_1$. Since p_1 and q_1 are irreducible, it follows that p_1 and q_1 are associates. Writing $q_1 = u p_1$ for some unit u, we can cancel p_1 from each of the above factorizations, yielding

$$p_2 p_3 \cdots p_m = u q_2 q_3 \cdots q_n.$$

Continuing in this way we conclude that $m = n$ and that each p_i has one of the q_i as an associate, and vice versa. Therefore, these two factorizations of x are equivalent. This completes the proof.

Fundamental to the proof of the above theorem is the idea that in a principal ideal domain an ascending chain of ideals cannot go on forever. Once again, the idea is that the union of the chain is itself an ideal, a principal ideal, whose generator must of course be in one of the ideals in the chain, and so, the chain stops ascending at that ideal. The general property that requires any ascending chain of ideals to terminate is now called the *ascending chain condition*.

The crucial importance of this *ascending chain condition* in ring theory was first recognized and developed by Emmy Noether. To honor both her and her work, the rings in which the ascending chain condition holds are now called *Noetherian rings*. These rings are the subject of Chapter 13. In particular, we should point out that we have just seen above that

principal ideal domains are Noetherian.

Euclidean Domains

It is really quite extraordinary, but the integers do seem to have an almost endless supply of interesting properties that are worthy of generalization. Another of these fundamental properties is the one that is based on the familiar division algorithm. Euclid used this property of the integers to give his well-known algorithm for finding the greatest common divisor of two integers. Therefore, the rings which have such a division algorithm have been named after him.

Definition 10.2. *An integral domain D is called a* **Euclidean domain** *if there is a function Δ, called the degree function,*

$$\Delta : D \setminus \{0\} \longrightarrow \mathbf{Z}^+ \cup \{0\};$$

that is, Δ is a function on the nonzero elements of D that takes on values in the non-negative integers — such that

(1) *for $x, y \neq 0 \in D$: if $x|y$, then $\Delta(x) \leq \Delta(y)$;*
(2) *for $x, y \in D, x \neq 0$: there are elements $q, r \in D$ such that*

$$y = qx + r \text{ where } r = 0 \text{ or } \Delta(r) < \Delta(x).$$

Example 4

Let us show that the ring of integers $\mathbf{Z}$ is a Euclidean domain. First, we must choose a degree function Δ. The obvious candidate is $\Delta(x) = |x|$ — that is, the absolute value function. Since the absolute value function takes integers to non-negative integers, all we have to do is show that Δ satisfies conditions (1) and (2).

(1) Let $x, y \neq 0 \in \mathbf{Z}$. If $x|y$, then $y = rx$ for some $r \neq 0 \in \mathbf{Z}$, and $|y| = |r| \cdot |x| \geq |x|$. Thus, $\Delta(x) \leq \Delta(y)$.
(2) Let $x, y \in \mathbf{Z}$ where $x \neq 0$. Then, by the division algorithm for the integers — that is, by ordinary long division — we can find integers q and r such that $y = qx + r$, where $0 \leq r < x$. Thus, $r = 0$ or $\Delta(r) < \Delta(x)$.

Example 5
Let K be a field. We will see in Problem 10.1 that the polynomial ring $K[x]$ is a Euclidean domain.

Example 6
The ring $\mathbf{Z}[i]$ of *Gaussian integers* is a Euclidean domain (see Problem 10.3).

We now wish to show that every Euclidean domain is a principal ideal domain. Thus, by Theorem 10.1, every Euclidean domain is a unique factorization domain.

Theorem 10.2. *Every Euclidean domain is a principal ideal domain.*

Proof
Let I be an ideal of a Euclidean domain D and let Δ be the degree function of D. We must show that I is a principal ideal. If $I = (0)$, then I is a principal ideal. Otherwise, $I \neq (0)$ and the set

$$\Delta(I) = \{\Delta(x) \mid x \neq 0 \in I\}$$

is a non-empty set of non-negative integers and, as such, has a smallest integer value. Let x be an element of I that takes on that smallest value — that is, $\Delta(x) \leq \Delta(y)$ for all $y \neq 0 \in I$. We claim that $I = (x)$.

Clearly, $(x) \subseteq I$. On the other hand, let $y \in I$. Then, $y = qx + r$ for some $q, r \in D$, where $r = 0$ or $\Delta(r) < \Delta(x)$. But, since $x, y \in I$, $r = y - qx \in I$, so $\Delta(r) \geq \Delta(x)$ by the minimality of x. Thus we conclude that $r = 0$, and so $y = qx$. Therefore, $y \in (x)$ and $I \subseteq (x)$, as desired. Hence, $I = (x)$ and I is a principal ideal, as claimed. This completes the proof.

We can summarize the relationship between unique factorization domains, principal ideal domains, and Euclidean domains as follows:

$$\text{Euclidean domain} \quad \Longrightarrow \quad \text{PID} \quad \Longrightarrow \quad \text{UFD}.$$

However, no two of these concepts are equivalent — that is, there are PIDs that are not Euclidean, and there are UFDs that are not PIDs. We will see, for example, in the next chapter that the polynomial ring $\mathbf{Z}[x, y]$ is a UFD; however, as we know, $\mathbf{Z}[x, y]$ is not a PID since (x, y) is not a principal ideal. It is considerably harder to give an example of a PID that is not Euclidean; this is because in order to show that a ring in *not* Euclidean, you must show that there is no function Δ *whatsoever* that works. One example of such a ring, however, is $\mathbf{Z}[\frac{1+\sqrt{-19}}{2}]$.

Problems

10.1   Let K be a field. Prove that $K[x]$ is a Euclidean domain. (Hint: it is this example that motivates our use of the term *degree function* for the function Δ in the definition of Euclidean domain.)

10.2   We will see in the next chapter that $\mathbf{Z}[x]$ is a unique factorization domain. Show, however, that $\mathbf{Z}[x]$ is *not* a principal ideal domain. (Hint: consider the ideal $I = \{2m + xf \mid m \in \mathbf{Z},\ f \in \mathbf{Z}[x]\}$; don't forget to show that I is an ideal.)

10.3   The ring $\mathbf{Z}[i]$ of *Gaussian integers* is defined to be the subring $\{a + bi \mid a, b \in \mathbf{Z}\}$ of the complex numbers $\mathbf{C}$. Show that $\mathbf{Z}[i]$ is a Euclidean domain by taking the *norm* function

$$N(a + bi) = a^2 + b^2$$

as your degree function Δ.
        (Hint: for property (2) let $x, y \in \mathbf{Z}[i]$; in order to find the quotient q and remainder r, first write $y/x = u + vi \in \mathbf{C}$, which can be done since $\mathbf{C}$ is a field, and then write $u = u_1 + u_2$ and $v = v_1 + v_2$ where $u_1, v_1 \in \mathbf{Z}$ and $|u_2| \le 1/2$ and $|v_2| \le 1/2$.)

10.4   It follows from Problem 10.3 that $\mathbf{Z}[i]$ is a principal ideal domain. Prove this directly by showing that any ideal I of $\mathbf{Z}[i]$ is principal. (Hint: use the norm function N from Problem 10.3 to pick an element $z \ne 0 \in I$ such that $N(z)$ is as small as possible; then show that the elements of the ideal (z) are just the vertices of an infinite family of squares that cover the entire complex plane; finally, show that $I = (z)$ by showing that if there were an element in $I \setminus (z)$, this would contradict the minimality of $N(z)$.)

10.5   Euclid used the division algorithm to find the greatest common divisor of two integers. We now call his process the *Euclidean algorithm*. For example, in order to find gcd(42, 30), we use the division algorithm to write $42 = 1 \cdot 30 + 12$, then we use the division algorithm again to write $30 = 2 \cdot 12 + 6$, and again to write $12 = 2 \cdot 6 + 0$, at which point we conclude that gcd(42, 30) = 6.
        We know from Problem 9.10 in that any pair of elements in $\mathbf{Z}[i]$ have a greatest common divisor. Find the greatest common divisor of $9 - 3i$ and $7 - 7i$. (Hint: here is a trick that will get you started on the

Euclidean algorithm:

$$\frac{9-3i}{7-7i} = \frac{9-3i}{7-7i} \cdot \frac{7+7i}{7+7i} = \frac{84+42i}{98} = \frac{6}{7} + \frac{3}{7}i;$$

and the nearest Gaussian integer to this quotient is $1 + 0i$.)

10.6   The Euclidean algorithm described in Problem 10.5 also allows us to
express the greatest common divisor of two elements as a linear
combination of the two elements. For example, $\gcd(42, 30) = 6$, and
we can express the greatest common divisor 6 in the form $r \cdot 42 + s \cdot 30$
by working backward along our computations in Problem 10.5 to get
$6 = 30 - 2 \cdot 12 = 30 - 2(42 - 1 \cdot 30) = 3 \cdot 30 - 2 \cdot 42$, thus expressing 6
as a linear combination of 30 and 42.
   Express $\gcd (9 - 3i, 7 - 7i)$ in the ring $\mathbf{Z}[i]$ as a linear combination of
$9 - 3i$ and $7 - 7i$.

   The two notions discussed in Problems 10.5 and 10.6 give rise to two
more generalizations of the integers! An integral domain in which
every pair of elements has a greatest common divisor is called a
GCD-domain. An integral domain satisfying the stronger condition
that the greatest common divisor of any pair of elements can always be
written as a linear combination of the two elements is called a **Bézout
domain**.

10.7   Show that $\mathbf{Z}[\sqrt{-2}] = \{a + b\sqrt{-2} \mid a, b \in \mathbf{Z}\}$ is a Euclidean domain.
(Hint: see Problem 10.3.)

10.8   Show that $\mathbf{Z}[\sqrt{5}] = \{a + b\sqrt{5} \mid a, b \in \mathbf{Z}\}$ is not a UFD, and, hence,
not a Euclidean domain. (Hint: in order to show that elements are
irreducible you will need to use the *norm* function
$N(a + b\sqrt{5}) = a^2 - 5b^2$.)

10.9   Find an irreducible element of $\mathbf{Z}[\sqrt{5}]$ that is not prime.

10.10   Show that $\mathbf{Z}[\frac{1+\sqrt{5}}{2}]$ is a Euclidean domain. (Hint: you can use the
norm of Problem 10.8.)

10.11   Find an element in the polynomial ring $\mathbf{R}[x^3, xy, y^3]$ that is irreducible
but not prime.

10.12   Let D be a principal ideal domain. Prove that, for an element $x \neq 0 \in D$,

x is irreducible if and only if (x) is maximal.

10.13   We know that $\mathbf{Z}[\sqrt{-5}]$ is not a PID (since, by Example 3 in Chapter 9, it is not a UFD). Show directly that $\mathbf{Z}[\sqrt{-5}]$ is not a PID by finding an ideal of $\mathbf{Z}[\sqrt{-5}]$ that is not principal.

10.14   Let D be a unique factorization domain. Prove that if every nonzero prime ideal of D is maximal, then D is a principal ideal domain.

10.15   A beautiful theorem due to Fermat says that any odd prime p can be written as a sum of two squares if and only if $p \equiv 1 \pmod 4$. Moreover, this representation as a sum of two squares is unique. So primes of the form $4k + 1$ are the sum of two squares, whereas primes of the form $4k + 3$ are not. For example, $137 = 4^2 + 11^2$, whereas 139 cannot be written as a sum of two squares. In this problem we will use the fact that $\mathbf{Z}[i]$ is a UFD to prove this remarkable theorem, which is known as *Fermat's two square theorem.*

   First, show that no prime of the form $4k + 3$ can be a sum of two squares. (Hint: show that a square is necessarily of the form $4k$ or $4k + 1$.)

   Now, let p be a prime of the form $4k + 1$. Then, make use of the fact from number theory that in this case -1 is a *quadratic residue* of p — that is, there is an integer a such that $a^2 \equiv -1 \pmod p$. (In fact, using Wilson's theorem, it is easy to show that $a = (\frac{p-1}{2})!$ works for this.)

   Next, show that p is *not* an irreducible element of $\mathbf{Z}[i]$. (Hint: $p | a^2 + 1 = (a + i)(a - i)$.)

   Then show that p is a sum of two squares by writing $p = (a + bi)(c + di)$, where neither factor is a unit, and by using the norm function. In fact, you should note that $c = a$ and $d = -b$, which, for example, explains why $137 = 4^2 + 11^2$, because $137 = (4 + 11i)(4 - 11i)$.

11

Polynomial Rings

We now focus our attention on a single ring, the ring $R[x]$ of polynomials in a single indeterminate x with coefficients from a ring R. We call this the *polynomial ring over R*. There are many questions to be asked about the ring $R[x]$. What are the units of this ring? What are the zero-divisors? What are its maximal ideals, its prime ideals? More generally, and perhaps more interestingly, what is the nature of the relationship between the ring R and the ring $R[x]$?

We have already seen in Problem 3.14 that if R is an integral domain, then so is $R[x]$. The famous *Hilbert basis theorem* — recall if you will our discussion of this theorem in Chapter 1 and see our proof, yet to come, in Chapter 13 — has this same form: if R is Noetherian, then so too is $R[x]$. Thus, a very typical sort of question for us to be asking will be: what specific ring-theoretic properties does the polynomial ring $R[x]$ inherit from its coefficient ring R? So, for example, polynomial rings inherit both the property of being an integral domain and the property of being a Noetherian ring from their coefficient rings.

Units in $R[x]$

Our first result in this chapter concerns the question of how to recognize whether or not a given polynomial f is a unit in $R[x]$. The answer is quite simple and involves, as it should, merely looking at the coefficients of f. Before stating and proving the relevant theorem, we do however, need to know a general fact concerning units, a fact which is a slight generalization of Problem 5.6. In any ring

the sum of a unit element and a nilpotent element is a unit.

In order to prove this fact, let u be a unit and let r be a nilpotent element of a ring R. We shall show that $u + r$ is a unit. Since r is nilpotent, $r^n = 0$ for some positive integer n. The inverse of $u + r$, then, is

$$u^{-1} - u^{-2}r + u^{-3}r^2 - \cdots + (-1)^{n-1}u^{-n}r^{n-1}.$$

We can verify this either by multiplying this expression by $u + r$ to get 1 (using the fact that $r^n = 0$), or by using long division to divide $u + r$

into 1 to get the above expression (observing that any subsequent terms in the quotient turn out to be 0 because they contain r^n).

We can now present the theorem that tells us how to recognize units in the polynomial ring $R[x]$.

Theorem 11.1. *Let R be a ring. An element $f \in R[x]$,*

$$f = a_0 + a_1 x + a_2 x^2 + \cdots + a_n x^n,$$

is a unit in $R[x]$ if and only if

$$a_0 \text{ is a unit of } R$$

and all of the other coefficients

$$a_1, a_2, \ldots, a_n \text{ are nilpotent in } R.$$

Proof

First, assume that a_0 is a unit and that $a_1, a_2, \ldots, a_n$ are nilpotent. We must show that f is a unit of $R[x]$. Since a_0 is a unit of R, it is also a unit of $R[x]$. Furthermore, since a_1 is nilpotent in R, $a_1 x$ is nilpotent in $R[x]$. So, by the fact proved above, $a_0 + a_1 x$ is a unit of $R[x]$. Similarly, $a_2 x^2$ is nilpotent in $R[x]$, and so, $(a_0 + a_1 x) + a_2 x^2$ is a unit of $R[x]$. Continuing in this fashion, we conclude that f is a unit of $R[x]$.

Conversely, we now assume f is a unit of $R[x]$. Then we define, for *each* prime ideal P of R, the natural homomorphism ϕ_P, where

$$\phi_P : R[x] \longrightarrow (R/P)[x]$$

is given by

$$\phi_P (a_0 + a_1 x + a_2 x^2 + \cdots + a_n x^n)$$
$$= (a_0 + P) + (a_1 + P)x + (a_2 + P)x^2 + \cdots + (a_n + P)x^n.$$

Now, since we are assuming that f is a unit of $R[x]$, we can write $fg = 1$ for some $g \in R[x]$. Then $\phi_P(f)\, \phi_P(g) = \phi_P(fg) = \phi_P(1) = 1$ for each prime P, so $\phi_P(f)$ is a unit in $(R/P)[x]$. But, by Theorem 3.1, R/P is an integral domain, so $(R/P)[x]$ is also an integral domain, by Problem 3.14. Thus, the polynomial $\phi_P(f)$ must have degree 0, since otherwise we could not have $\phi_P(f)\, \phi_P(g) = 1$.

This means that $\phi_P(f)$ is a constant polynomial, and hence a unit in R/P for each prime P. We conclude that $a_0 \notin P$, but that $a_i \in P$ for $i > 0$. But, this is true for *all* primes P. So, a_0 is not contained in *any* maximal ideal; hence, by Theorem 5.1, a_0 is a unit in R. Similarly, each

a_i, for $i > 0$, lies in *every* prime ideal P; hence, by Theorem 5.2, each a_i, for $i > 0$, is nilpotent in R. This completes the proof of the theorem.

Zero-Divisors in $R[x]$

By definition, an element f in $R[x]$ is a zero-divisor if there is a polynomial $g \neq 0$ in $R[x]$ such that $fg = 0$. Our next theorem says that if f is a zero-divisor, then not only is there a nonzero polynomial g that does the job, but there is a nonzero constant that does the same thing!

Theorem 11.2 (McCoy, 1942). *Let R be a ring. A polynomial f in $R[x]$ is a zero-divisor if and only if there is a nonzero element $a \in R$ such that $af = 0$.*

Proof

Suppose that $f \in R[x]$ is a zero-divisor. Write $f = a_0 + a_1 x + \cdots + a_n x^n$, and let

$$g = b_0 + b_1 x + \cdots + b_m x^m$$

be a nonzero polynomial of *least* degree m such that $fg = 0$.

First, we observe that $a_n b_m = 0$. Therefore, the polynomial $a_n g$ has degree less than g. But, $(a_n g) f = 0$, since $fg = 0$. By the minimality of g, we conclude that $a_n g = 0$.

Now, repeat the foregoing argument for a_{n-1}, then for a_{n-2}, and so on. Thus each of the coefficients a_i of f has the property that $a_i g = 0$, and so, in particular, for each coefficient, $a_i b_m = 0$. Therefore, $b_m f = 0$, and b_m is the constant we are seeking. (In fact, it actually follows that $m = 0$ and that $g = b_0$.) This proves the necessity of our condition for f to be a zero-divisor, and its sufficiency is of course obvious.

Unique Factorization and $R[x]$

Our next goal is to prove the important result that if R is a unique factorization domain, then so too is the polynomial ring $R[x]$. In other words, the polynomial ring inherits unique factorization from its coefficient ring. The proof we develop is due to Gauss. In fact, some authors even use the term *Gaussian domains* for unique factorization domains. The proof is quite simple but does require careful preparation.

Recall from Problem 9.10 that in a unique factorization domain any pair of elements has a greatest common divisor. We now show that, in a *unique factorization domain,*

any finite set of elements has a greatest common divisor.

We give a proof by induction on n, the number of elements in the set. For $n = 2$, this is just Problem 9.10. Assume that the statement is true for any set consisting of fewer than n elements in a unique factorization domain. Let $\{a_1, \ldots, a_n\}$ be a set of n elements. Then, by the induction hypothesis, the set $\{a_1, \ldots, a_{n-1}\}$ has a greatest common divisor; call it d_1. The two elements d_1 and a_n also have a greatest common divisor; call it d. We claim that d is the greatest common divisor of the elements $a_1, \ldots, a_n$.

Since d divides d_1, d also divides each of $a_1, \ldots, a_{n-1}$, and it divides a_n as well. Therefore, d is a common divisor of $a_1, \ldots, a_n$. Suppose that e is also a common divisor of $a_1, \ldots, a_n$. Then e divides $a_1, \ldots, a_{n-1}$, so e divides d_1. But, e also divides a_n, so e divides d. Therefore, d is the greatest common divisor of $a_1, \ldots, a_n$. We conclude that in a unique factorization domain any finite set of elements has a greatest common divisor, as claimed.

We also need the following term.

Definition 11.1. *A polynomial is said to be* **primitive** *if the only divisors of all its coefficients are units.*

For example, in a unique factorization domain it is possible to write any polynomial f as $f = cg$, where c is a constant and g is primitive. In order to see this, let R be a unique factorization domain, and let $f \in R[x]$. Simply let c be the greatest common divisor of the coefficients of f. Then, if $f = a_0 + a_1 x + \cdots + a_n x^n$, we can let

$$g = \frac{a_0}{c} + \left(\frac{a_1}{c}\right) x + \cdots + \left(\frac{a_n}{c}\right) x^n,$$

which is clearly a primitive polynomial, and $f = cg$. This leads us to another useful term:

Definition 11.2. *Let R be a unique factorization domain, and let $f \in R[x]$. The* **content** *of the polynomial f is a greatest common divisor of its coefficients. We denote the content of f by $c(f)$.*

For example, a polynomial is primitive if its content is a unit. Of course, the content of a polynomial is unique only up to multiplication by units. The justification for saying *the* content is historical, since this notion was first applied to the case where the coefficient ring is the integers, and in this case the content was taken to be the greatest common positive divisor.

It is easy to see that we can factor a polynomial f into a product of irreducible elements if the coefficient ring R is a unique factorization domain. We simply write $f = cg$, where c is a constant and g is a primitive polynomial. Since R is a unique factorization domain, c can be uniquely factored as a product of irreducible elements in R, but these elements are also irreducible in $R[x]$. Finally, since the degree of the polynomial g is finite, any nontrivial factorization of g and its subsequent factors must eventually result in irreducible factors of positive degree. However, in order to show the uniqueness of such a factorization, we need the following three lemmas.

Gauss's lemma. *The product of two primitive polynomials with coefficients from a unique factorization domain is also a primitive polynomial.*

Proof
Let f and g be primitive polynomials over a unique factorization domain R and write.

$$f = a_0 + a_1 x + \cdots + a_n x^n \text{ and } g = b_0 + b_1 x + \cdots + b_m x^m.$$

We give a proof by contradiction. Assume that if the product fg is not primitive, then there is an irreducible non-unit d in R which divides all the coefficients of fg.

Since f and g are each primitive, the coefficients of each have no common divisors other than units. In particular, the element d cannot divide all the coefficients of either f or g. Let a_i be the first coefficient of f — moving from left to right — which is not divisible by d, and let b_j be the first coefficient of g which is not divisible by d. Since d divides *all* of the coefficients of fg, d divides the coefficient of x^{i+j} in the product fg, but this coefficient is just

$$(\cdots + a_{i-1} b_{j+1} + a_i b_j + a_{i+1} b_{j-1} + \cdots).$$

Now, d divides each term to the left of $a_i b_j$ by the choice of i (since $d|a_0, d|a_1, \ldots, d|a_{i-1}$), and d divides each term to the right of $a_i b_j$ by the choice of j (since $d|b_{j-1}, d|b_{j-2}, \ldots, d|b_0$). But d divides the entire expression as well, so d must also divide $a_i b_j$. But, since d is irreducible, it is prime by Problem 9.6, so d divides a_i or d divides b_j, and this is a contradiction. We therefore conclude that fg is primitive. This completes the proof of Gauss's lemma.

Lemma 1. *Let R be a unique factorization domain, and let F be its quotient field. Let f and g be primitive polynomials in $R[x]$. If f and g are associates in $F[x]$, then they are associates in $R[x]$.*

Proof

Since f and g are associates in $F[x]$, we can write $f = ug$, where the polynomial u is a unit in $F[x]$ — that is, $u = a_0 + a_1x + \cdots + a_nx^n$ where, by Theorem 11.1, a_0 is a unit in F and the other coefficients $a_1, a_2, \ldots, a_n$ are nilpotent in F. But, the only nilpotent element in the field F is 0, so $a_1 = 0, a_2 = 0, \ldots, a_n = 0$ and $u = a_0$ — that is, u is a unit and is in F. So $u \neq 0 \in F$, and we can write $u = a/b$ for some $a, b \in R$. Thus, $bf = ag$; but, f and g are primitive polynomials in $R[x]$, so b and a both represent the content of the same polynomial in $R[x]$ and, therefore, b and a are associates in R. We conclude that f and g are associates in $R[x]$, as claimed. This completes the proof of this lemma.

Lemma 2. *Let R be a unique factorization domain, and let F be its quotient field. If f is a nonconstant irreducible polynomial in $R[x]$, then f is also irreducible in $F[x]$.*

Proof

We give a proof by contradiction. Suppose, by way of contradiction, that $f = gh$ for some polynomials $g, h \in F[x]$ such that the degree of each of g and h is less than that of f — that is, suppose that we can properly factor f in $F[x]$. We show that this leads to a proper factorization of f in $R[x]$.

Now let $a, b \in R$ be such that $ag, bh \in R[x]$; for example, a and b could be the product of all the denominators of the coefficients of g and h, respectively. We can write $ag = cg'$ and $bh = dh'$, where $c, d \in R$ and g' and h' are primitive polynomials in $R[x]$. Then, $abf = abgh = cdg'h'$. By Gauss's lemma, $g'h'$ is a primitive polynomial in $R[x]$. But f is also primitive in $R[x]$, since it is irreducible in $R[x]$; hence f and $g'h'$ are associates in $R[x]$. This contradicts our assumption that f is irreducible in $R[x]$. We therefore conclude from this contradiction that f is irreducible in $F[x]$. This completes the proof of this final lemma.

At last, we can now deal with the *uniqueness* of factorization for polynomials over a unique factorization domain.

Theorem 11.3. *If R is a unique factorization domain, then so is the polynomial ring $R[x]$.*

Proof

Let $f \in R[x]$ and suppose that the polynomial f has two factorizations

$$f = c_1c_2 \cdots c_j g_1 g_2 \cdots g_r = d_1 d_2 \cdots d_k h_1 h_2 \cdots h_s,$$

where the c_i and d_i are irreducible constants in R and the g_i and h_i are nonconstant, irreducible polynomials in $R[x]$. We must show that these two factorizations are equivalent.

Since the g_i and h_i are irreducible, they are primitive in $R[x]$. Thus, by Gauss's lemma, the products $g_1 g_2 \cdots g_r$ and $h_1 h_2 \cdots h_s$ are primitive in $R[x]$. So, $c_1 c_2 \cdots c_j$ and $d_1 d_2 \cdots d_k$ are associates in R. But, R is a unique factorization domain, so the c_i and the d_i are the same up to units in R.

Also, then, $g_1 g_2 \cdots g_r$ and $h_1 h_2 \cdots h_s$ are associates in $R[x]$, and hence, are associates in $F[x]$. But $F[x]$ is a unique factorization domain (it is in fact a Euclidean domain by Problem 10.1), and by Lemma 2 the g_i and h_i are irreducible in $F[x]$, so the g_i and the h_i must pair off into associates in $F[x]$. Finally, by Lemma 1, they are also associates in $R[x]$, and the factorization is unique. This completes the proof.

Problems

11.1 Prove that if R is a unique factorization domain, then so is $R[x_1, \ldots, x_n]$, the ring of polynomials in n indeterminates $x_1, \ldots, x_n$.

11.2 Is the polynomial $5 + 6x + 3x^2$ a unit in $Z/(12)[x]$? (Here, for convenience, we are writing elements of $Z/(12)$ as integers mod 12 rather than as cosets.)

11.3 Is the polynomial $2 + 6x + 3x^2$ a zero-divisor in $Z/(12)[x]$? (Here, again, we are writing elements of $Z/(12)$ as integers mod 12 rather than as cosets.)

11.4 Find the $\gcd(x^5 - 2x^4 - x + 2, x^4 - 4x^3 + 5x^2 - 4x + 4)$ in $Q[x]$ both by using the Euclidean algorithm (see Problem 10.5) and by factoring each polynomial into irreducible factors.

11.5 Let R be a unique factorization domain, and let $f, g \in R[x]$. Prove that $c(fg) = c(f)c(g)$.

11.6 A polynomial is called *monic* if the coefficient of its highest power is 1. Prove the following (which is also sometimes called Gauss's lemma). Let f be a monic polynomial in $Z[x]$ and let g and h be monic polynomials in $Q[x]$ such that $f = gh$; then $g, h \in Z[x]$.

11.7 Let K be a field. Find all of the *maximal* ideals of $K[x]$.

11.8 Let K be a field. Find all of the *prime* ideals of $K[x]$.

11.9 Let I be an ideal of a ring R, and define $I[x]$ to be the set of all polynomials whose coefficients are in I. Prove that

$$R[x]/I[x] \cong (R/I)[x].$$

11.10 Let P be an ideal of a ring R. Prove that P is prime in R if and only if $P[x]$ is prime in $R[x]$.

11.11 Let R be a ring. Prove that an element $f \in R[x]$ is nilpotent in $R[x]$ if and only if all of its coefficients are nilpotent in R. (Hint: use Problem 11.10 and Theorem 5.2.)

11.12 Is the polynomial $4 + 6x + 3x^2$ nilpotent in $Z/(12)[x]$? (Here, once again, we are writing elements of $Z/(12)$ as integers mod 12 rather than as cosets.)

11.13   Let M be a maximal ideal of a ring R. Prove that $M[x]$ is *not* maximal in $R[x]$.

11.14   Let M be a maximal ideal in a ring R. Prove that any prime ideal of $R[x]$ properly containing $M[x]$ is a maximal ideal of $R[x]$.

   In this case, we would say that the prime ideal $M[x]$ has **dimension** 1, since the ideal $M[x]$ is "sitting" one step below a maximal ideal and not more than that below any other prime ideal. (An older, highly descriptive, term for this is "depth." In Chapter 14 we will define the *dimension* of a ring; the **dimension** of an ideal I of a ring R, then, is given by dim $I = $ dim R/I).

11.15   Let R be a ring. Show that a maximal ideal of $R[x]$ need not contain x. (Hint: show that $1 - x$ is contained in a maximal ideal.)

11.16   Let K be a field and let $x_1, \ldots, x_n$ be indeterminates. Prove, for any i, $1 \le i \le n$, that $(x_1, \ldots, x_i)$ is a prime ideal of $K[x_1, \ldots, x_n]$. (Hint: use Theorem 3.1.) In particular, then, we say that

$$(0) \subset (x_1) \subset (x_1, x_2) \subset \cdots \subset (x_1, x_2, \ldots, x_n)$$

is a *chain* of prime ideals of length n in this ring.

12

Power Series Rings

Power Series

Surely among the most beautiful and useful ideas in all of mathematics is a notion that emerged during the eighteenth century that familiar and important functions such as the sine function, the cosine function, and the exponential function can be represented by infinite series:

$$e^x = 1 + x + \frac{x^2}{2!} + \frac{x^3}{3!} + \frac{x^4}{4!} + \cdots,$$

$$\sin x = x - \frac{x^3}{3!} + \frac{x^5}{5!} - \frac{x^7}{7!} + \cdots,$$

$$\cos x = 1 - \frac{x^2}{2!} + \frac{x^4}{4!} - \frac{x^6}{6!} + \cdots.$$

For example, the above infinite series representations of $\sin x$, $\cos x$, and e^x allow us to extend these real-valued functions to the complex plane and, by using the infinite series representations of these functions, we are thus able to uncover the deep connection between the two trigonometric functions and the exponential function that is revealed to us by *Euler's formula*:

$$e^{iz} = \cos z + i \sin z.$$

Of course, when dealing with infinite series, among the main issues that arise are questions involving convergence. In this chapter on power series, however, we will be working with infinite series, but without worrying at all about such matters as their convergence. In other words, we will consider infinite series *only* from an algebraic point of view. In fact, we even change their name to *power series* in order to emphasize that change in point of view.

Definition 12.1. *A **power series** in a single indeterminate x over a ring R is an expression of the form*

$$f = a_0 + a_1 x + a_2 x^2 + a_3 x^3 + \cdots,$$

where the coefficients a_i come from the ring R.

We can define addition and multiplication of power series in an obvious way by imitating those same operations for polynomials. For example,

$$\left(1 + x + \frac{x^2}{2} + \frac{x^3}{6} + \frac{x^4}{24} + \cdots\right)\left(x - \frac{x^3}{6} + \frac{x^5}{120} + \cdots\right)$$

$$= x + x^2 + \frac{x^3}{3} - \frac{x^5}{30} + \cdots .$$

With these natural operations of addition and multiplication, the set of all power series over a given ring R is a ring, called the *power series ring* over R. We denote this ring by $R[[x]]$.

The zero element of a power series ring is of course $0 + 0x + 0x^2 + 0x^3 + \cdots$, which we usually write as 0, and the multiplicative identity is $1 + 0x + 0x^2 + 0x^3 + \cdots$, which we usually write as 1.

Power series look, and behave, like extremely long polynomials, so it is not surprising that power series rings are quite like polynomial rings in many ways. However, they are surprisingly different in other respects. Our main theme in this chapter will be to compare and contrast these two types of rings. To understand the interplay between the rings R and $R[x]$ on the one hand, and the rings R and $R[[x]]$ on the other, is to understand much about the nature of commutative ring theory.

Units in $R[[x]]$

Our first result concerns the question of how to recognize whether a given power series f is a unit in $R[[x]]$. The corresponding answer to that question for polynomial rings — Theorem 11.1 — was quite simple, but for power series the answer is even simpler. In fact, recognizing units in $R[[x]]$ is about as simple as it could possibly be.

Theorem 12.1. *Let R be a ring. A power series*

$$f = a_0 + a_1 x + a_2 x^2 + \cdots$$

is a unit in R[[x]] if and only if

$$a_0 \text{ is a unit in } R.$$

Proof

Let f be a unit in $R[[x]]$. Then $fg = 1$ for some power series $g \in R[[x]]$. Writing $g = b_0 + b_1 x + b_2 x^2 + \cdots$, we see that $a_0 b_0 = 1$, and so, a_0 is a unit in R.

Conversely, suppose that a_0 is a unit in R. The idea is, quite simply, to divide f into 1 by long division. Since a_0 is a unit, this process can be carried out indefinitely, giving — term by term — the inverse of f. This argument, albeit valid, may be less than convincing, however. We therefore restate the argument somewhat more formally — that is, inductively.

Define $b_0 = a_0^{-1}$. Now, assume that $b_0, b_1, \ldots, b_n$ have been defined — that is, the first $n + 1$ terms of the inverse g have been found. The coefficient of x^{n+1} in the product fg is supposed to be 0, so we require that

$$a_0 b_{n+1} + a_1 b_n + \cdots + a_{n+1} b_0 = 0.$$

Solving this equation for b_{n+1} we get

$$b_{n+1} = -a_0^{-1}(a_1 b_n + \cdots + a_{n+1} b_0).$$

We therefore use this equation to define b_{n+1} in terms of the previously defined terms $b_0, \ldots, b_n$. Thus, we define a power series g inductively, and easily see that $fg = 1$. This completes the proof.

Zero-Divisors in $R[[x]]$

The result above that tells us how to recognize units in $R[[x]]$ certainly gives the impression that working with power series rings might be more pleasant than working with polynomial rings. Sometimes it is, and sometimes it isn't. As a matter of fact, when it comes to recognizing zero-divisors, the situation is actually reversed.

Because of Theorem 11.2 (McCoy's theorem), it is easy for us to recognize zero-divisors in polynomial rings. We only need to look within the coefficient ring for potential elements to multiply by a given polynomial to test whether it is a zero-divisor. The following example, which we state as a theorem for emphasis, shows that no such pleasant result is remotely possible for power series. In fact, as the theorem shows, it is actually possible for a power series to be a zero-divisor, and yet to have 1 as one of its coefficients!

Theorem 12.2 (Fields, 1971). *There exist a ring R and a power series of the form $f = a_0 + x \in R[[x]]$ such that f is a zero-divisor in $R[[x]]$.*

Proof

Let K be a field, let $y, z_0, z_1, z_2, \ldots$ be indeterminates, and form the polynomial ring $K[y, z_0, z_1, z_2, \ldots]$. Then

$$I = (yz_0, \; z_0 + yz_1, \; z_1 + yz_2, \; z_2 + yz_3, \ldots)$$

is an ideal in $K[y, z_0, z_1, z_2, \ldots]$.

Now, let

$$R = K[y, z_0, z_1, z_2, \ldots]/I,$$

and let

$$f = (y + I) + x \in R[[x]].$$

We claim that the ring R and the power series f have the desired property: the power series f is a zero-divisor in $R[[x]]$. For, if we let

$$g = (z_0 + I) + (z_1 + I)x + (z_2 + I)x^2 + \ldots,$$

then we easily see that $fg = 0$. This completes the proof.

In particular, this theorem shows us that Theorem 11.2 fails completely for power series. Our only hope of recapturing the spirit of McCoy's theorem for power series is by placing suitable restrictions on the ring R. One such restriction we could make is to require that R be Noetherian (see the next chapter). Another is to require that R be a what is called a *reduced* ring:

Definition 12.2. *A ring R is a **reduced ring** if R has no nonzero nilpotent elements.*

In other words, by Theorem 5.2, a ring is reduced if the intersection of all of its prime ideals is (0) — that is, if the nilradical $\mathcal{N}$ is (0). For example, any integral domain is reduced. The following theorem tells us that the product of two power series over a *reduced* ring is 0 *only* if all of the individual products of their coefficients are 0.

Theorem 12.3. *Let R be a reduced ring, and let $f, g \in R[[x]]$. If $f = a_0 + a_1 x + a_2 x^2 + \ldots$ and $g = b_0 + b_1 x + b_2 x^2 + \cdots$, then*

$$fg = 0 \text{ if and only if } a_i b_j = 0 \text{ for all } i \text{ and } j.$$

Proof
Assume that $fg = 0$. First, we show that $a_0b_j = 0$ for all j. We give a proof by induction. Clearly, $a_0b_0 = 0$. Now assume that $a_0b_0 = a_0b_1 = \cdots = a_0b_n = 0$. Since $fg = 0$, the coefficient of the x^{n+1} term in the product fg is 0 — that is,

$$a_0b_{n+1} + a_1b_n + \cdots + a_{n+1}b_0 = 0.$$

Thus,

$$0 = a_0(a_0b_{n+1} + a_1b_n + \cdots + a_{n+1}b_0) = a_0^2b_{n+1},$$

and so $(a_0b_{n+1})^2 = 0$. But, since R is a reduced ring, the only nilpotent element in R is 0, and so $a_0b_{n+1} = 0$, as desired. Hence, by induction, $a_0b_j = 0$ for all j.

Now, since $fg = 0$ and $a_0b_j = 0$ for all j, it follows that

$$(a_1x + a_2x^2 + \cdots)g = 0.$$

Dividing by x, we get

$$(a_1 + a_2x + a_3x^2 + \cdots)g = 0,$$

and we can repeat the argument given above to show that $a_1b_j = 0$ for all j. This can then be repeated for a_2, a_3, and so on. Thus, we conclude that if $fg = 0$, then $a_ib_j = 0$ for all i and j.

Since the converse is obvious, this completes the proof.

Maximal Ideals of $R[[x]]$

We now turn to the question of recognizing maximal ideals in $R[[x]]$. In Problem 11.15 we saw that a maximal ideal of the polynomial ring $R[x]$ need not contain x. But, for power series rings, a maximal ideal of $R[[x]]$ *always* contains x. In order to characterize the maximal ideals in $R[[x]]$, the following notation will be convenient.

Let I be an ideal of a ring R. The set of all power series whose constant terms are elements of I — that is, all $f = a_0 + a_1x + a_2x^2 + \cdots \in R[[x]]$ such that $a_0 \in I$ — is an ideal of $R[[x]]$. The routine verification of this fact is left to you (see Problem 12.7). We denote this ideal by

$$I + (x) = \{f \in R[[x]] \mid f_0 \in I\}.$$

Note that $I + (x)$ is not the *sum* of two ideals of $R[[x]]$ since I is not an ideal of $R[[x]]$.

The maximal ideals of $R[[x]]$ are now easy to identify.

Theorem 12.4. *Let R be a ring. The maximal ideals of R[[x]] are all ideals of the form*

$$M + (x),$$

where M is a maximal ideal of R.

Proof

First, we show that if M is a maximal ideal of R, then $M + (x)$ is a maximal ideal in $R[[x]]$. Let

$$\phi : R[[x]] \longrightarrow R$$

be the homomorphism defined by $\phi(f) = a_0$ for $f = a_0 + a_1 x + a_2 x^2 + \cdots \in R[[x]]$. As a homomorphism which is clearly onto, ϕ gives a one-to-one, order-preserving correspondence between the ideals of R and the ideals of $R[[x]]$ that contain the kernel of ϕ — that is, ideals that contain (x).

Since M is a maximal ideal in R, it follows that $\phi^{-1}(M)$ is maximal in $R[[x]]$. But, $\phi^{-1}(M) = M + (x)$, so $M + (x)$ is maximal in $R[[x]]$.

Conversely, if N is a maximal ideal in $R[[x]]$, then we must produce a maximal ideal M of R such that $N = M + (x)$. The natural candidate is

$$M = \{a_0 \mid f = a_0 + a_1 x + a_2 x^2 + \cdots \in N\};$$

that is, $M = \phi(N)$.

Clearly, M is an ideal. It is not equal to R, because if $1 \in M$, then N contains some power series $f = a_0 + a_1 x + a_2 x^2 + \cdots$ such that $a_0 = 1$. But, then, f is a unit, by Theorem 12.1, in a maximal ideal N, which is impossible. Thus, M is not equal to R.

Next, we show that M is not contained in any other proper ideal. Let $a \in R \setminus M$. We must show that $(M, a) = R$. First, $a \in R[[x]] \setminus N$, so that $(N, a) = R[[x]]$ by the maximality of N. We can therefore write $1 = f + ag$ for some $f = a_0 + a_1 x + a_2 x^2 + \cdots \in N$ and $g = b_0 + b_1 x + b_2 x^2 + \cdots \in R[[x]]$. But, this means that $1 = a_0 + ab_0$, and so, $1 \in (M, a)$. Hence, $(M, a) = R$, after all, and M is a maximal ideal of R, as claimed.

Finally, we observe that $N \subseteq M + (x) \subset R[[x]]$, but N is maximal, so $N = M + (x)$, as desired. This completes the proof.

It may have occurred to you that the notation (M, a) is ambiguous. Is it the ideal generated by M and a in the ring R or in the ring $R[[x]]$? For example, in the above proof, when we wrote (M, a) we meant the ideal in R. If the context does not make it clear, then we can write $(M, a)R$ or $(M, a)R[[x]]$ in order to make the necessary distinction.

Unique Factorization and $R[[x]]$

What properties can a power series ring $R[[x]]$ inherit from its coefficient ring R? One result of this kind — one that was true for polynomial rings — is certainly true for power series rings: namely, the property of being an integral domain passes from the coefficient ring to the power series ring (see Problem 12.1).

Similarly, at this stage you may well be expecting us to prove that if R is a unique factorization domain, then so is $R[[x]]$ — a fact that was true for polynomial rings. Surprisingly, and very disappointingly, this is simply not true. This had been an important open question in commutative ring theory for many years, and was finally resolved — albeit negatively — in 1961 by Pierre Samuel, who demonstrated that R could have unique factorization without $R[[x]]$ having unique factorization.

Just as they always do when faced with disappointing truths, mathematicians looked for restrictions that could be placed on rings so that the desired result on unique factorization for power series would still hold true. Happily, it turns out that if the stronger restriction of being a principal ideal domain is placed on the coefficient ring R, then $R[[x]]$ will be a unique factorization domain. For the proof of this fact you will have to wait until the next chapter.

Problems

12.1   Prove that if R is an integral domain, then so is $R[[x]]$; hence, prove that if R is an integral domain, then so is $R[[x_1, \ldots, x_n]]$.

12.2   Let R be a ring. Show that $R[x]$ is a subring of $R[[x]]$.

12.3   Is the power series $5 + 2x + 2x^2 + 2x^3 + \cdots$ a unit in $\mathbf{Z}/(12)[[x]]$? (Here, for convenience, we are writing elements of $\mathbf{Z}/(12)$ as integers mod 12 rather than as cosets.)

12.4   Show that $\mathbf{Z}/(12)$ is *not* a reduced ring. Is the power series $2 + 3x + 2x^2 + 3x^3 + 2x^4 + 3x^5 + \cdots$ a zero-divisor in $\mathbf{Z}/(12)[[x]]$? (Here, again, we are writing elements of $\mathbf{Z}/(12)$ as integers mod 12 rather than as cosets.)

12.5   Let K be a field. Find all of the *maximal* ideals of $K[[x]]$.

12.6   Let K be a field. Find all of the *prime* ideals of $K[[x]]$.

12.7   Let I be an ideal of a ring R. Show that the set

$$I + (x) = \left\{ f = a_0 + a_1 x + a_2 x^2 + \cdots \in R[[x]] \mid a_0 \in I \right\}$$

is an ideal of $R[[x]]$.

12.8   Let R be a ring. Prove that an ideal P is prime in R if and only if $P + (x)$ is prime in $R[[x]]$.

12.9   Let I be an ideal of a ring R, and define $I[[x]]$ to be the set of all power series whose coefficients lie in I. Prove that

$$R[[x]]/I[[x]] \cong (R/I)[[x]].$$

12.10   Let R be a ring. Prove that an ideal P is prime in R if and only if $P[[x]]$ is prime in $R[[x]]$.

12.11   Let R be a ring. Prove that if an element $f \in R[[x]]$ is nilpotent in $R[[x]]$, then all of its coefficients are nilpotent in R.

12.12   Let $y_0, y_1, y_2, \ldots$ be indeterminates and form the polynomial ring $\mathbf{Q}[y_0, y_1, y_2, \ldots]$. Then, let n be an integer greater than 1 and let I be the ideal $(y_0^n, y_1^n, y_2^n, \ldots)$. Finally, let R be the quotient ring $\mathbf{Q}[y_0, y_1, y_2, \ldots]/I$.

Show that the power series $f = a_0 + a_1x + a_2x^2 + \cdots$ — where $a_i = y_i + I$ for $i \geq 0$ — is *not* nilpotent in $R[[x]]$, but that $a_i^n = 0$ for each i. (This example is also due to Fields.)

12.13   Let M be a maximal ideal of a ring R. Prove that $M[[x]]$ is not maximal in $R[[x]]$.

12.14   Let K be a field. Show that $(x_1, \ldots, x_n)$ is the unique maximal ideal of $K[[x_1, \ldots, x_n]]$; hence, $K[[x_1, \ldots, x_n]]$ is a local ring.

13

Noetherian Rings

The Hilbert Basis Theorem

It was Emmy Noether who first recognized the importance of the *ascending chain condition*. It is therefore entirely fitting that the rings that satisfy this condition are named after her.

Definition 13.1. *A ring R is said to satisfy the* **ascending chain condition** *if every ascending chain of ideals*

$$I_1 \subseteq I_2 \subseteq I_3 \subseteq \cdots$$

terminates — that is, there is an integer k such that

$$I_1 \subseteq I_2 \subseteq \cdots \subseteq I_k = I_{k+1} = I_{k+2} = \cdots .$$

For example, we saw in Chapter 10 that principal ideal domains satisfy the ascending chain condition. In particular, this is true because any ideal in a principal ideal domain is generated by a single element; hence the union of an ascending chain of ideals is itself an ideal that is generated by a single element, which, in turn, must be in one of the ideals in the chain, forcing the chain to terminate at this ideal. More generally, and for exactly the same reason, if R is a ring such that any ideal of R is generated by a *finite* number of elements, then R satisfies the ascending chain condition (see Problem 13.3).

An ideal which is generated by a finite number of elements is called a **finitely generated ideal**. Since, conversely to our conclusion above, it is also true that any ring that satisfies the ascending chain condition has the property that all of its ideals are finitely generated (again, see Problem 13.3), we give our definition of Noetherian rings in the following form:

Definition 13.2. *A ring R is called* **Noetherian** *if it satisfies the ascending chain condition — or, equivalently, if every ideal of R is finitely generated.*

Noetherian rings are of tremendous importance, not only for their own sake and historically, but also because many of the rings that turn up naturally, for example in algebraic geometry, are Noetherian.

We are now almost set to prove the famous Hilbert basis theorem, first mentioned in Chapter 1. In order to do this, however, it will be useful to introduce yet one more term concerning polynomials. Let R be a ring, and let $f \in R[x]$. The **leading coefficient** of the polynomial f is the coefficient of the highest power term in f. Admittedly, since we have been writing polynomials in ascending order, the term "leading" coefficient may seem a bit strange for the coefficient of the last term. However, since it is perhaps more common to write polynomials in descending order, the term "leading" coefficient is well entrenched in the literature.

Let I be an ideal of $R[x]$. Let J be the set of *all* of the leading coefficients of *all* of the polynomials in I. We now show that J is an ideal of R. Let a_m and b_n be two elements of J — that is, a_m and b_n are the leading coefficients of two polynomials $f = a_0 + a_1 x + \cdots + a_m x^m$ and $g = b_0 + b_1 x + \cdots + b_n x^n$ in I of degrees m and n, respectively. We may as well assume that $m \le n$. Then $x^{n-m} f - g \in I$ and has leading coefficient $a_m - b_n$. Therefore, $a_m - b_n \in J$.

Next, let a_m be an element of J — that is, a_m is the leading coefficient of a polynomial f in I of degree m. Let $r \in R$. Then $rf \in I$, so ra_m is the leading coefficient of a polynomial in I, and so $ra_m \in J$. Thus the set J of leading coefficients from I is an ideal of R.

Theorem 13.1 (Hilbert basis theorem). *If R is a Noetherian ring, then so is the polynomial ring $R[x]$.*

Proof
Let I be an ideal in $R[x]$. We must show that I is finitely generated. For each non-negative integer n, let J_n be the ideal of R consisting of the leading coefficients of polynomials in I whose degrees are less than or equal to n. Then

$$J_0 \subseteq J_1 \subseteq J_2 \subseteq \cdots$$

is an ascending chain of ideals in the Noetherian ring R and, as such, must terminate, say, at J_k. So

$$J_0 \subseteq J_1 \subseteq \cdots \subseteq J_k = J_{k+1} = \cdots .$$

Now, J_i is finitely generated for each $i = 1, 2, \ldots, k$. Therefore, for each i, let $f_{i_1}, f_{i_2}, \ldots, f_{i_r}$ be polynomials in $R[x]$ whose leading coefficients generate J_i in R. This gives us a finite set of polynomials, and it is not at all hard to see that this finite set of polynomials generates the ideal I. This completes the proof.

The following corollary is itself also often referred to as the Hilbert basis theorem and is an immediate consequence of Theorem 13.1. In fact, in Chapter 15 when we make use of the Hilbert basis theorem it is really this corollary we will have in mind.

Corollary 1. *Let K be a field, and let $x_1, x_2, \ldots, x_n$ be n indeterminates. Then $K[x_1, x_2, \ldots, x_n]$ is Noetherian.*

A Hilbert Basis Theorem for $R[[x]]$

We now would like to show that there is an analogue of the Hilbert basis theorem for power series — namely, if a ring R is Noetherian, then so is the power series ring $R[[x]]$. In order to prove this, however, we need to have two additional theorems in place first.

Recall that in order for a ring to be Noetherian our definition says that *all* of its ideals must be finitely generated. The first theorem tells us that in fact we only have to check the prime ideals in order to verify that a ring is Noetherian!

Theorem 13.2 (I. S. Cohen). *If every prime ideal of a ring R is finitely generated, then R is Noetherian.*

Proof

We give a proof by contradiction. Suppose, by way of contradiction, that R is not Noetherian. If we let S be the set of ideals of R that are not finitely generated, then, by assumption, S is not empty. Furthermore, since all prime ideals are finitely generated, S does not contain any prime ideals. We are preparing to use Zorn's lemma, that is, Theorem 4.1.

Let $\{I_\lambda\}$ be a chain of ideals in S. The union of this chain is also an ideal in S: The union is an ideal, and if the union were finitely generated, then its generators would necessarily be in one of the ideals I_λ, making that particular I_λ finitely generated, which it is not. Thus, by Zorn's lemma, there is a maximal element P in S. We shall reach a contradiction by showing that P is a prime ideal.

Assume that P is not prime. Then, there are elements $a, b \notin P$ such that $ab \in P$. Thus, the ideal (P, a) is strictly larger than P and, hence, by the maximality of P, must be finitely generated. We can therefore write

$$(P, a) = (p_1 + r_1 a, \ p_2 + r_2 a, \ \ldots, \ p_m + r_m a),$$

where $p_i \in P$ and $r_i \in R$ for $i = 1, 2, \ldots, m$.

Now, let $I = \{x \in R \mid xa \in P\}$. Then, I is an ideal and, since $I \supseteq (P, b)$, and (P, b) is strictly larger than P, I is also finitely generated. Thus, we can write

$$I = (i_1, i_2, \ldots, i_n).$$

We claim that

$$P = (p_1, \ldots, p_m, i_1 a, \ldots, i_n a).$$

In order to show this, we let $p \in P$. Then, $p \in (P, a)$, and so, $p = s_1(p_1 + r_1 a) + \cdots + s_m(p_m + r_m a)$ for some $s_i \in R$, $i = 1, 2, \ldots, m$. But, $s_1 p_1 + \cdots + s_m p_m \in P$, so $(s_1 r_1 + \cdots + s_m r_m)a \in P$. By the definition of I, this means that $s_1 r_1 + \cdots + s_m r_m \in I$ and can be expressed in terms of $i_1, \ldots, i_n$. Thus, P is finitely generated, as claimed, which is a contradiction. We conclude that P is prime. This in turn contradicts the hypothesis that all prime ideals are finitely generated. Thus, R is Noetherian. This completes the proof.

Our next theorem now gives us a way to decide if a given prime ideal in a power series ring is indeed finitely generated. We shall make use of the following fact (see Problem 13.1). Let $\phi : R \to S$ be an onto homomorphism between two rings R and S. Then, if an ideal I of R is finitely generated, its homomorphic image $\phi(I)$ is also finitely generated.

Theorem 13.3 (Kaplansky). *Let R be a ring, and let P be a prime ideal of the power series ring $R[[x]]$. Let $P_0 = \{a_0 \mid f = a_0 + a_1 x + \cdots \in P\}$. Then P is finitely generated in $R[[x]]$ if and only if P_0 is finitely generated in R. In fact, if P_0 is generated by k elements, then P can be generated by $k + 1$, or k, elements according to whether the element x is, or is not, in P.*

Proof
Let $\phi : R[[x]] \to R$ be the homomorphism given by $\phi(f) = a_0$, for $f = a_0 + a_1 x + \cdots$. Then $\phi(P) = P_0$. It follows from Problem 13.1 that if P is finitely generated, then so is its homomorphic image P_0.

Conversely, suppose that P_0 is finitely generated, and write $P_0 = (a_1, a_2, \ldots, a_k)$. We first consider the easy case in which $x \in P$. Then $P = (a_1, \ldots, a_k, x)$, and so P is generated by $k + 1$ elements.

Next, we consider the case in which $x \notin P$. Let $f_1, f_2, \ldots, f_k$ be power series in P that have $a_1, a_2, \ldots, a_k$ respectively as their constant terms. We will show that $P = (f_1, \ldots, f_k)$. Let $g = b_0 + b_1 x + \cdots \in P$. We must show that $g \in (f_1, \ldots, f_k)$. Since $b_0 \in P_0$, we can write

$$b_0 = r_1 a_1 + \cdots + r_k a_k,$$

for some $r_i \in R$. It follows that

$$g - (r_1 f_1 + \cdots + r_k f_k) = xg',$$

for some $g' \in R[[x]]$. Thus $xg' \in P$, but $x \notin P$, so $g' \in P$, since P is prime. Similarly,

$$g' - (s_1 f_1 + \cdots + s_k f_k) = xg'',$$

for some $s_i \in R$ and some $g'' \in P$.

Continuing in this way, we generate, term by term, power series $g_1, \ldots, g_k \in R[[x]]$ by writing

$$g_i = r_i + s_i x + \cdots$$

for $i = 1, 2, \ldots, k$. It is easy to see that $g = g_1 f_1 + \cdots + g_k f_k$. Thus, $g \in (f_1, \ldots, f_k)$. We conclude that $P = (f_1, \ldots, f_k)$, and so P is finitely generated, as claimed. This completes the proof.

An immediate consequence of Theorems 13.3 and 13.2 is the following power series analogue of the Hilbert basis theorem.

Theorem 13.4. *If a ring R is Noetherian, then so too is the power series ring $R[[x]]$.*

Unique Factorization in $R[[x]]$

The foregoing theorem of Kaplansky, Theorem 13.3, also allows us to return to the problem we discussed in the previous chapter of showing that $R[[x]]$ is a unique factorization domain when R is a principal ideal domain — that is, when R is the ultimate among Noetherian rings, since, in this case, not only is every ideal of R finitely generated, but every ideal of R is generated by a single element. This important result will be easy to demonstrate once we provide a new and useful characterization of unique factorization domains. This characterization of unique factorization domains is also due to Kaplansky.

We shall make use of another fact which we leave for you to show: namely, that if an element in an integral domain is expressible as a product $p_1 p_2 \cdots p_k$ of prime elements, then that expression is unique (see Problem 13.2).

Theorem 13.5 (Kaplansky). *An integral domain D is a unique factorization domain if and only if every nonzero prime ideal of D contains a nonzero prime element.*

Proof

Assume that D is a unique factorization domain. Let P be a nonzero prime ideal, and let $a \neq 0 \in P$. Write the element a as a product of irreducible elements, $a = p_1 \cdots p_k$. Then $p_1 \cdots p_k \in P$, so $p_i \in P$ for some $i = 1, \ldots, k$, by Problem 3.1. But, then p_i is also a prime element, since p_i is irreducible and irreducible elements are also prime in unique factorization domains (see Problem 9.6). Hence P contains a nonzero prime element p_i.

Conversely, assume every nonzero prime ideal of D contains a nonzero prime element. Because of Problem 13.2, in order to show that D is a unique factorization domain, we need only show that each non-unit is a product of prime elements. Let A be the set of all products of nonzero prime elements of D. Let $d \neq 0 \in D$ be a non-unit. It suffices to show that $d \in A$.

Assume that $d \notin A$. We claim that $(d) \cap A = \emptyset$. For, if not, let $rd \in A$ with $r \in D$. We can then write $rd = p_1 \cdots p_k$ as a product of primes p_i. So $p_1 | r$ or $p_1 | d$. Suppose that $p_1 | d$; then we write $d = p_1 d_1$ for some $d_1 \in D$; otherwise, $p_1 | r$, and we write $r = p_1 r_1$ for some $r_1 \in D$. Then, since D is an integral domain, $rd_1 = p_2 \cdots p_k$ in the first case, or $r_1 d = p_2 \cdots p_k$ in the second. Continuing in this way, we end up, after k steps, with $r_i d_j = 1$ for some i and j, and d has been written as a product of prime elements, and so, d is in the set A. But, we assumed that $d \notin A$. From this contradiction we conclude that $(d) \cap A = \emptyset$.

We now use Zorn's lemma to enlarge the ideal (d) to an ideal that is maximal with respect to *missing* A. Let S be the set consisting of all ideals of D that contain (d) but are disjoint from A. The set S is non-empty since the ideal (d) is itself in S. The union of any chain of ideals in S is also clearly in S. So, by Zorn's lemma, S has a maximal element P. (We are not claiming that P is a maximal ideal, just that P is maximal among those ideals that miss A and contain (d).)

We shall show that P is a prime ideal. Let $ab \in P$, and suppose, by way of contradiction, that neither a nor b is in P. Since the ideal (P, a) is strictly larger than P, it must meet A. Thus there is an element a_1 of A such that $a_1 = p_1 + r_1 a$ for some $p_1 \in P$ and $r_1 \in D$. Similarly, there is an element $a_2 \in A$ such that $a_2 = p_2 + r_2 b$ for some $p_2 \in P$ and $r_2 \in D$. Now,

$$a_1 a_2 = (p_1 + r_1 a)(p_2 + r_2 b) \in P \cap A,$$

since $ab \in P$. This contradicts the fact that $P \cap A = \emptyset$. Thus P is prime.

But this in turn contradicts the hypothesis of the theorem, since P cannot contain a prime element (being disjoint from A). Thus our assumption that $d \notin A$ was wrong. Hence any non-unit d can be written as a product of prime elements. So, by Problem 13.2, D is a unique factorization domain. This completes the proof.

This characterization of unique factorization domains allows us now to give a condition on a ring R which guarantees that the power series ring $R[[x]]$ is a unique factorization domain.

Theorem 13.6. *If R is a principal ideal domain, then the power series ring $R[[x]]$ is a unique factorization domain.*

Proof
By the preceding theorem, we only have to show that every nonzero prime ideal P of $R[[x]]$ contains a nonzero prime element. If $x \in P$, then P contains the prime element x. If $x \notin P$, then, by Theorem 13.3, the ideal P can be generated by the same number of elements that it takes to generate the ideal P_0. But, R is a principal ideal domain, so P_0 is a principal ideal. Therefore, P is also a principal ideal — that is, $P = (f)$ for some $f \neq 0 \in R[[x]]$. But, then, f is a nonzero prime element in P.

Problems

13.1   Let $f : R \to S$ be an onto homomorphism between two rings R and S. Show that if an ideal I of R is finitely generated, then its homomorphic image $f(I)$ is also finitely generated.

13.2   Recall that an element p of a ring R is *prime* if the ideal (p) is prime in R. In other words, p is prime if $p|ab$ implies that $p|a$ or $p|b$. Show that if an element in an integral domain is expressible as a product $p_1 p_2 \cdots p_k$ of prime elements, then that expression is unique.

13.3   Prove that a ring R satisfies the ascending chain condition if and only if every ideal of R is finitely generated.

13.4   Prove that any homomorphic image of a Noetherian ring is Noetherian — that is, prove that if I is an ideal of a Noetherian ring R, then R/I is also Noetherian.

13.5   Show that the ring $\mathbf{Z}[i]$ of Gaussian integers is Noetherian.

13.6   Let I be an ideal of a ring R such that every ascending chain of ideals of R contained in I terminates. Prove that if R/I is Noetherian, then R is Noetherian.

13.7   Let R be a Noetherian ring and let $f : R \to R$ be an onto homomorphism. Prove that f is one-to-one.

13.8   Let $R = \mathbf{Q}[x_1, x_2, x_3, \dots]$ be the ring of polynomials in infinitely many indeterminates $x_1, x_2, \dots$. Show that R is not Noetherian. In particular, find both an ascending chain of ideals of R that does not terminate and an ideal of R that is not finitely generated.

13.9   Show that a subring of a Noetherian ring need not be Noetherian.

13.10   Let R be the ring of *eventually constant* sequences of 0s and 1s — that is, sequences consisting of only the elements 0 and 1 such that, after a certain point, the terms are either *all* 0 or *all* 1, and addition and multiplication are done componentwise. For example, the sequences $(0, 1, 0, 1, 1, 1, 1, 1, \dots)$ and $(1, 0, 1, 0, 0, 0, 0, 0, \dots)$ are two elements of R, their sum is the identity element $(1, 1, 1, 1, 1, 1, 1, 1, \dots)$, and their product is the zero element $(0, 0, 0, 0, 0, 0, 0, 0, \dots)$.

    Show that R is not Noetherian. In particular, find both an ascending chain of ideals of R that does not terminate and an ideal of R that is not finitely generated.

13.11   Show that the ring $C([0, 1])$ of continuous real-valued functions on the closed interval $[0, 1]$ is not Noetherian. (Hint: for a closed interval F in $[0, 1]$, define $I = \{f \in C([0, 1]) \mid f(F) = 0\}$.)

13.12   Let R be a ring. If $R[x]$ is Noetherian, is R necessarily Noetherian?

13.13   Let R be a ring. If $R[[x]]$ is Noetherian, is R necessarily Noetherian?

13.14   Let T be a multiplicative system in a ring R. Prove that if R is Noetherian, then so too is the localization R_T.

13.15   Let R be a ring. If R_P is Noetherian for all prime ideals P, is R necessarily Noetherian? In other words, is the property of being Noetherian a *local* property (see Problem 6.20)?

13.16   Let D be an integral domain. Prove that D is a unique factorization domain if and only if every irreducible element is prime and the principal ideals satisfy the ascending chain condition.

13.17   Let R be a Noetherian ring. For a power series $f \in R[[x]]$, define the *content ideal* $C(f)$ of f to be the ideal of R generated by the coefficients of f. Prove that a power series f in $R[[x]]$ is nilpotent if and only if there is an integer k such that $f_i^k = 0$ for all coefficients f_i of f (see Problems 12.11 and 12.12).

13.18   Let R be a Noetherian ring. Prove that an element $f \in R[[x]]$ is nilpotent in $R[[x]]$ if and only if all of its coefficients are nilpotent in R.

14

Dimension

The Krull Dimension of a Ring

The notion of dimension is of obvious importance in the study of geometry, but the concept itself is somewhat elusive. It is all well and good for us to say that n-dimensional space has n dimensions, but what exactly can be said that accurately describes the essential difference between, say, a one-dimensional line and a two-dimensional plane once we have seen the remarkable *space filling curve* discovered by Hilbert? By the early part of the twentieth century a topological definition of the notion of dimension had been developed with sufficient precision to give genuine meaning to statements concerning dimension that previously were merely intuitive, such as claiming that space has three dimensions.

Commutative ring theory evolved in no small part to establish a link between algebra and geometry, and so it was certainly natural to look for an algebraic analogue of the topological notion of dimension. The following definition of the algebraic notion of dimension was first introduced by Wolfgang Krull in 1937, and is now so commonly viewed as the "correct" definition that the official terminology *Krull dimension* — which is used in part to honor Krull — is often shortened to simply *dimension*.

Definition 14.1. *The **Krull dimension** of a ring R is the supremum, or least upper bound, of the lengths of chains of prime ideals in R. We write* **dim R** *for the dimension of R. The **length** of a chain of distinct prime ideals*

$$P_0 \subset P_1 \subset P_2 \subset \cdots \subset P_k$$

is said to be k.

Let's look at several examples.

Example 1
If K is a field, then $\dim K = 0$. This is easy to see since, by Problem 2.20, the ideal $P_0 = (0)$ is the only proper ideal in K. Hence, since it is also a prime ideal, it forms the longest chain as well.

In Problem 14.1 you are asked to show that there is a converse to this statement, namely, that if R is an integral domain, and dim $R = 0$, then R is a field.

Example 2
The ring $\mathbf{Z}$ of integers has dimension 1. The longest chains of prime ideals in $\mathbf{Z}$ all look like $(0) \subset (p)$, where p is a prime integer.

Example 3
Let K be a field. Then, the polynomial ring $K[x]$ has dimension 1. So, in particular,

$$\dim K[x] = \dim K + 1.$$

For instance, the chain $(0) \subset (x)$ is a chain of length 1. No longer chain is possible simply because $K[x]$ is a principal ideal domain. In fact, the prime ideals of $K[x]$ are just the ideal (0) together with all of the maximal ideals, which are the principal ideals generated by the irreducible polynomials.

As you can see, both $\mathbf{Z}$ and $K[x]$ have dimension 1 precisely because they are principal ideal domains (see Problem 14.2).

Example 3 is a special case of an extremely important result about the dimension of $K[x_1, x_2, \ldots, x_n]$, the polynomial ring in n indeterminates over a field K. The chain of ideals

$$(0) \subset (x_1) \subset (x_1, x_2) \subset \cdots \subset (x_1, x_2, \ldots, x_n)$$

is a chain of distinct prime ideals of length n in this ring. The relatively easy task of showing that these ideals are in fact prime ideals was as an exercise in Chapter 11 (see Problem 11.16).

In fact, it turns out that there are no longer chains of prime ideals in this ring; in other words, the polynomial ring $K[x_1, x_2, \ldots, x_n]$ has dimension n; that is,

$$\dim K[x_1, x_2, \ldots, x_n] = \dim K + n.$$

Proving this important result will be one of our primary objectives in this chapter.

The Dimension of Polynomial Rings
In this section we continue with our investigation of the nature of the relationship between a ring R and its polynomial ring $R[x]$ in the

context of dimension. Recall from Problem 11.9 that for an ideal I, we define $I[x]$ to be the set of all polynomials whose coefficients are in I. It is easy to see that $I[x] \cap R = I$, since the "constant" polynomials in $I[x]$ are those polynomials $f = a_0$ where $a_0 \in I$.

Since $I[x]$ is the kernel of the natural homomorphism from $R[x]$ to $(R/I)[x]$, it follows that

$$R[x]/I[x] \cong (R/I)[x].$$

In particular, then, an ideal P is a prime ideal if and only if $P[x]$ is a prime ideal, by Theorem 3.1 and Problem 3.14.

Now, the ideal in $R[x]$ generated by I, which we denote by $IR[x]$, is clearly contained in the ideal $I[x]$. In fact, these two ideals are equal; that is, $IR[x] = I[x]$ (see Problem 14.3).

A very useful fact concerning the polynomial ring $R[x]$ is that if T is a multiplicative system of R, then

$$(R[x])_T \cong R_T[x].$$

Verification of this is left as an exercise in Problem 14.4.

For example, if R is an integral domain, then we can let T be the multiplicative system of all nonzero elements of R. In this case, then $(R[x])_T$ is just $K[x]$, where K is the quotient field of R. So, by Theorem 6.1, there is a one-to-one correspondence between the prime ideals in $K[x]$ and the prime ideals of $R[x]$ which are disjoint from T. But, a prime ideal Q in $R[x]$ is disjoint from T if and only if $Q \cap R = (0)$. Thus, we conclude that there is a one-to-one correspondence between the prime ideals in $K[x]$ and the prime ideals of $R[x]$ whose intersection with R is (0).

An immediate consequence of this is the following theorem, which is of fundamental importance in the study of chains of prime ideals in polynomial rings. The theorem says simply that it is impossible to have a chain of three distinct prime ideals in a polynomial ring $R[x]$ whose intersections with R are all equal.

Theorem 14.1. *Let R be a ring, and let $Q_1 \supset Q_2 \supset Q_3$ be a chain of three distinct prime ideals in $R[x]$. Then, $Q_1 \cap R \neq Q_3 \cap R$.*

Proof
We immediately switch to the ring $(R/Q_3)[x]$. This gives us a chain of distinct prime ideals of $(R/Q_3)[x]$, namely,

$$Q_1/Q_3 \supset Q_2/Q_3 \supset Q_3/Q_3 = (0).$$

Assume, by way of contradiction, that $Q_1 \cap R = Q_3 \cap R$. This means that all three ideals Q_1/Q_3, Q_2/Q_3, and Q_3/Q_3 have intersection (0) with R/Q_3.

Since R/Q_3 is an integral domain, we can invoke our previous discussion, letting L represent the quotient field of R/Q_3, and so, by virtue of the previously described one-to-one correspondence, these three ideals whose intersection with R/Q_3 is (0) correspond to a chain of three distinct prime ideals in $L[x]$. But this is clearly impossible because, as we saw in Example 3, dim $L[x] = 1$. This completes the proof.

Theorem 14.1 allows us to give upper and lower bounds on the dimension of a polynomial ring $R[x]$ in terms of the dimension of R. The proof of this result is left as an exercise in Problem 14.6.

Theorem 14.2. *Let R be a ring, and let* dim $R = k$. *Then,*

$$k + 1 \leq \dim R[x] \leq 2k + 1.$$

Remarkably, for any k and any integer n within the bounds proscribed by this theorem, there is a ring R of dimension k such that the dimension of $R[x]$ is n. In other words, in the absence of additional conditions on the ring R, Theorem 14.2 tells us all that can be said about the dimension of $R[x]$. With that, we now turn toward our first main goal in this chapter: showing that, for a field K, the polynomial ring $K[x_1, x_2, \ldots, x_n]$ has dimension n. We begin with a convenient definition.

Definition 14.2. *Let P be a prime ideal of a ring R. The **height** of P — written **ht**(P) — is the supremum, or least upper bound, of the lengths of chains of prime ideals in R descending from P. Thus, if P has height k, there is a chain of distinct prime ideals*

$$P = P_0 \supset P_1 \supset P_2 \supset \cdots \supset P_k,$$

but none longer.

The height of a prime P is sometimes referred to as the *codimension* of P, or, in an older terminology, as the *rank* of P. But, the term *height* seems nicely descriptive of the situation. So, for example, minimal prime ideals have height 0, whereas the maximal ideals in the ring of integers $\mathbf{Z}$ or in a polynomial ring $K[x]$ over a field K all have height 1.

Next, we prove a useful intermediate result.

Theorem 14.3. *Let $x, x_2, \ldots, x_n$ be n indeterminates, where $n > 1$. Let R be a ring. Let Q be a prime ideal of $R[x]$, and let the ideal P of R be the prime* *ideal* $P = Q \cap R$. *If* $Q \supset P[x]$ — *that is, if* Q *and* $P[x]$ *are distinct primes of* $R[x]$ — *then*

$$\text{ht}(Q) = \text{ht}(P[x]) + 1,$$

and

$$\text{ht}(Q[x_2, \ldots, x_n]) = \text{ht}(P[x, x_2, \ldots, x_n]) + 1.$$

Proof
If $\text{ht}(P) = \infty$, then both results are obvious. So we assume that $\text{ht}(P)$ is finite and use induction on $\text{ht}(P)$ to prove the first result.

First, then, we suppose that $\text{ht}(P) = 0$. It follows that $\text{ht}(P[x]) = 0$, for, if not, we have

$$Q \supset P[x] \supset Q_1$$

for some prime ideal Q_1 of $R[x]$. But then, by Theorem 14.1, $Q \cap R \neq Q_1 \cap R$. Therefore, $Q_1 \cap R$ is a prime ideal properly contained in P, contradicting that $\text{ht}(P) = 0$. Hence, $\text{ht}(P[x]) = 0$.

Similarly, we can see that $\text{ht}(Q) = 1$. This is because, for any prime ideal $Q_1 \subset Q$, $Q_1 \cap R$ must be P, since $\text{ht}(P) = 0$. Thus, a chain of distinct prime ideals $Q \supset Q_1 \supset Q_2$ is impossible, by Theorem 14.1. (In fact, what this really tells us is that $P[x]$ is the unique prime ideal properly contained in Q.) Thus, in this case, $\text{ht}(Q) = \text{ht}(P[x]) + 1$, as desired.

Now, let $\text{ht}(P) = m$ for $m > 0$, and assume that this first result is true for $k < m$. We claim that, for any prime $Q_1 \subset Q$, $\text{ht}(Q_1) \leq \text{ht}(P[x])$. In order to prove this claim, let $P_1 = Q_1 \cap R$. If $P_1 = P$, then $P[x] \subseteq Q_1 \subset Q$ forces $Q_1 = P[x]$ by Theorem 14.1, and so $\text{ht}(Q_1) = \text{ht}(P[x])$, which means, of course, that $\text{ht}(Q_1) \leq \text{ht}(P[x])$ as well.

On the other hand, if $P_1 \subset P$, then either $Q_1 = P_1[x]$, or $Q_1 \supset P_1[x]$. In the first case, of course,

$$\text{ht}(Q_1) = \text{ht}(P_1[x]) \leq \text{ht}(P[x]),$$

and in the latter case we can use induction (since we know that $\text{ht}(P_1) < m$) to get

$$\text{ht}(Q_1) = \text{ht}(P_1[x]) + 1 \leq \text{ht}(P[x]).$$

Thus, the claim is verified.

But it now follows immediately from this claim that $\text{ht}(Q) \leq \text{ht}(P[x]) + 1$. And, of course, since $Q \supset P[x]$, it has been obvious all along that $\text{ht}(Q) \geq \text{ht}(P[x]) + 1$. Thus $\text{ht}(Q) = \text{ht}(P[x]) + 1$. This completes the proof of the first result of the theorem.

Next, as we turn to the second result, let us briefly consider the ring $R[x, x_2, \ldots, x_n]$ as $R[x][x_2, \ldots, x_n]$, so that we can still view $Q \subset R[x]$ and $P = Q \cap R$. Then we can see that

$$P[x_2, \ldots, x_n] = Q[x_2, \ldots, x_n] \cap R[x_2, \ldots, x_n].$$

This allows us to apply the first result of the theorem, since $Q[x_2, \ldots, x_n]$ is a prime ideal of $R[x_2, \ldots, x_n][x]$, and

$$Q[x_2, \ldots, x_n] \supset P[x_2, \ldots, x_n][x],$$

since $Q \supset P[x]$. Note that we are now considering the ring $R[x, x_2, \ldots, x_n]$ as $R[x_2, \ldots, x_n][x]$.

Therefore, by the first result, we have

$$\mathrm{ht}(Q[x_2, \ldots, x_n]) = \mathrm{ht}(P[x_2, \ldots, x_n][x]) + 1,$$

which is the second result, since $P[x_2, \ldots, x_n][x] = P[x, x_2, \ldots, x_n]$. This completes the proof of the theorem.

Now, we can use Theorem 14.3 to prove a major result:

Theorem 14.4. *Let R be a ring. Let Q be a prime ideal of the ring $R[x_1, x_2, \ldots, x_n]$, and let the ideal P of R be the prime ideal $P = Q \cap R$. Then*

$$\mathrm{ht}(Q) = \mathrm{ht}(P[x_1, \ldots, x_n]) + \mathrm{ht}\big(Q/P[x_1, \ldots, x_n]\big)$$

$$\leq \mathrm{ht}(P[x_1, \ldots, x_n]) + n.$$

Proof
We give a proof by induction on n. First, we do the case $n = 1$. If $Q = P[x_1]$, then we are done, since $\mathrm{ht}(Q) = \mathrm{ht}(P[x_1])$. If $Q \supset P[x_1]$, then, by Theorem 14.3, $\mathrm{ht}(Q) = \mathrm{ht}(P[x_1]) + 1$. But $\mathrm{ht}\big(Q/P[x_1]\big) = 1$ by Theorem 14.1, and so

$$\mathrm{ht}(Q) = \mathrm{ht}(P[x_1]) + \mathrm{ht}\big(Q/P[x_1]\big) \leq \mathrm{ht}(P[x_1]) + 1,$$

as desired.

Now we let $n > 1$, and assume that the theorem is true for $k < n$. Let $Q_1 = Q \cap R[x_1]$. Consider first the case where $Q_1 = P[x_1]$. This case follows immediately using induction in the ring $R[x_1][x_2, \ldots, x_n]$, where

$k = n - 1$, and we have

$$\mathrm{ht}(Q) = \mathrm{ht}(P[x_1][x_2, \ldots, x_n]) + \mathrm{ht}(Q/P[x_1][x_2, \ldots, x_n])$$

$$\leq \mathrm{ht}(P[x_1][x_2, \ldots, x_n]) + n - 1.$$

Next, consider the case where $Q_1 \supset P[x_1]$. In this case, the first thing to observe is that, by Theorem 14.3,

$$\mathrm{ht}(Q_1[x_2, \ldots, x_n]) = \mathrm{ht}(P[x_1, \ldots, x_n]) + 1.$$

Therefore, by induction, we can get

$$\mathrm{ht}(Q) = \mathrm{ht}(Q_1[x_2, \ldots, x_n]) + \mathrm{ht}(Q/Q_1[x_2, \ldots, x_n])$$

$$\leq \mathrm{ht}(Q_1[x_2, \ldots, x_n]) + n - 1$$

$$= \mathrm{ht}(P[x_1, \ldots, x_n]) + 1 + n - 1$$

$$= \mathrm{ht}(P[x_1, \ldots, x_n]) + n.$$

Thus we have shown, as desired, that $\mathrm{ht}(Q) \leq \mathrm{ht}(P[x_1, \ldots, x_n]) + n$.
In order to complete the proof, we still need to show that

$$\mathrm{ht}(Q) = \mathrm{ht}(P[x_1, \ldots, x_n]) + \mathrm{ht}(Q/P[x_1, \ldots, x_n]).$$

To do this, we first need to make the trivial observation that, since $Q_1 \supset P[x_1]$,

$$\mathrm{ht}(Q/P[x_1, \ldots, x_n]) \geq \mathrm{ht}(Q/Q_1[x_2, \ldots, x_n]) + 1,$$

a fact that will be used very shortly. We now simply repeat the argument used previously, so, once again, by induction we get

$$\mathrm{ht}(Q) = \mathrm{ht}(Q_1[x_2, \ldots, x_n]) + \mathrm{ht}(Q/Q_1[x_2, \ldots, x_n])$$

$$= \mathrm{ht}(P[x_1, \ldots, x_n]) + 1 + \mathrm{ht}(Q/Q_1[x_2, \ldots, x_n])$$

$$\leq \mathrm{ht}(P[x_1, \ldots, x_n]) + \mathrm{ht}(Q/P[x_1, \ldots, x_n]).$$

But, since it is obvious that

$$\mathrm{ht}(Q) \geq \mathrm{ht}(P[x_1, \ldots, x_n]) + \mathrm{ht}(Q/P[x_1, \ldots, x_n]),$$

it follows that

$$\mathrm{ht}(Q) = \mathrm{ht}(P[x_1, \ldots, x_n]) + \mathrm{ht}\big(Q/P[x_1, \ldots, x_n]\big),$$

as desired. This completes the proof of the theorem.

Theorem 14.4 is a truly remarkable theorem of Jim Brewer and Bill Heinzer. Its proof depends upon nothing more than the elementary Theorem 14.1, and yet it is powerful enough to prove many of the classical results in dimension theory. We now use it to attain what has been one of our primary goals in this chapter.

Theorem 14.5. *Let K be a field. Then*

$$\dim K[x_1, \ldots, x_n] = n.$$

Proof
By Problem 11.16, we know that

$$(0) \subset (x_1) \subset (x_1, x_2) \subset \cdots \subset (x_1, x_2, \ldots, x_n)$$

is a chain of distinct prime ideals of length n in this ring; hence, $\dim K[x_1, \ldots, x_n] \geq n$.

Now, suppose that Q is a prime ideal of the ring $K[x_1, x_2, \ldots, x_n]$. Then $Q \cap K = (0)$. (Otherwise, since K is a field, $Q \cap K = K$, and then $1 \in Q$.) So, by Theorem 14.4, $\mathrm{ht}(Q) \leq 0 + n$. Therefore, $\dim K[x_1, \ldots, x_n] \leq n$.

Hence, $\dim K[x_1, \ldots, x_n] = n$. This completes the proof of the theorem.

Since for a field K, $\dim K = 0$, the result in Theorem 14.5 has the form $\dim R[x_1, \ldots, x_n] = n + \dim R$. Theorem 14.4 can be used to prove that there are categories of rings other than fields where this same relationship holds between the dimension of a ring and the dimensions of the polynomial rings. We will now turn to the task of showing that Noetherian rings behave in this same way. This task, however, requires some preliminary groundwork. We will also find it necessary to invoke, without proof, one of the great landmark theorems of Noetherian ring theory, Krull's *principal ideal theorem*. We begin, as usual, with a definition.

Definition 14.3. *Let I be an ideal of a ring R. The* **radical** *of I — written $\sqrt{I}$ — is the following ideal of R:*

$$\sqrt{I} = \{a \in R \mid a^n \in I \text{ for some } n > 0\}.$$

It is easy to verify directly that $\sqrt{I}$ is, in fact, an ideal (see Problem 14.7). It is also easy to see this in another way, since the radical of I is the intersection of *all* prime ideals containing I. You are asked to use Zorn's lemma to prove this in Problem 14.8. Thus, for a radical ideal, we write

$$\sqrt{I} = \bigcap_{P \supseteq I} P.$$

This, of course, should remind you of Theorem 5.2, since $\sqrt{(0)}$ is nothing more than the *nilradical*. More generally, $\sqrt{I}/I$ is the nilradical of the ring R/I.

Also, recall from Problem 4.5 that every ideal I is contained in at least one prime ideal P which is *minimal over I* — that is, no prime ideal Q properly contained in P also contains I. Obviously, then, the radical of I is the intersection of all the *minimal* primes over I.

The radical of an ideal is a concept that comes into play in a very natural way when commutative algebra is used in the study of geometry. In particular, radicals of ideals are a central feature in Hilbert's fundamental theorem in this area, his famous *Nullstellensatz* of 1893, the theorem that forever linked geometry with modern algebra.

The radical of an ideal also has an important role to play in our current investigation of the dimension of polynomial rings over Noetherian rings, as we see in the next theorem.

Theorem 14.6. *Let R be a Noetherian ring, and let I be an ideal of R. Then, there are only finitely many primes minimal over I.*

Proof
Assume not. Then, by the ascending chain condition in R, we can further assume, without loss of generality, that the ideal I is maximal such that the statement in the theorem is false.

Now, $\sqrt{I}$ is the intersection of all the *minimal* primes over I. In particular, since this is not a finite intersection, $\sqrt{I}$ is not itself a prime ideal. So, we can let $a, b \in R$ be such that $ab \in \sqrt{I}$, but $a \notin \sqrt{I}$ and $b \notin \sqrt{I}$.

Therefore, by the maximality of I, there are only finitely many primes minimal over the ideal (I, a), and, similarly, only finitely many primes minimal over the ideal (I, b). Thus $\sqrt{(I, a)}$ is a *finite* intersection of the minimal primes over (I, a), and $\sqrt{(I, b)}$ is a *finite* intersection of the minimal primes over (I, b).

But, it is routine to check — this is left for you as an exercise in Problem 14.12 — that

$$\sqrt{I} = \sqrt{(I, a)} \cap \sqrt{(I, b)},$$

which means that $\sqrt{I}$ is a *finite* intersection of primes, and, hence, of minimal primes. Thus, there are only finitely many primes minimal over I, a contradiction. This completes the proof.

Our next theorem says that in a Noetherian ring any prime ideal can be thought of as a *minimal* prime in the sense that we can always find an ideal for it to be minimal over.

Theorem 14.7. *Let R be a Noetherian ring, and let P be a prime ideal of R. Let $ht(P) = n$. Then, there are elements $a_1, \ldots, a_n$ such that P is minimal over $(a_1, \ldots, a_n)$.*

Proof
Let $ht(P) = n$. We use induction on n. For $n = 0$, P itself is a minimal prime, and we can let the set of elements that P is minimal over be empty. So assume $n > 0$, and that the statement of the theorem is true for all primes having height less than n.

By Theorem 14.6, R has a finite number of minimal primes — that is, primes minimal over (0). Let $Q_1, \ldots, Q_k$ be these minimal primes of R. Since $ht(P) \geq 1$, $P \not\subseteq Q_i$, for any i. It then follows — this is Problem 14.13 — that

$$P \not\subseteq Q_1 \cup \cdots \cup Q_k.$$

So choose $a_1 \in P$, where $a_1 \notin Q_i$, for any i.

Now, in the ring $R/(a_1)$, consider the prime ideal $P/(a_1)$. We claim that $ht(P/(a_1)) \leq n - 1$. Suppose you have a chain of maximum length descending from $P/(a_1)$. Then this chain corresponds, under the inverse image of the natural homomorhism from R to $R/(a_1)$, to a chain of exactly the same length descending from P, each prime in the chain containing the ideal (a_1). But the bottom prime in this chain can't be a minimal prime because a_1 is contained in *no* minimal prime in R. Therefore, the chain descending from P can be extended, and the claim is verified.

So we can now use induction on the prime $P/(a_1)$, and there are $n - 1$ elements $a_2 + (a_1), \ldots, a_n + (a_1)$ such that $P/(a_1)$ is minimal over

$(a_2 + (a_1), \ldots, a_n + (a_1))$. Then, P is minimal over $(a_1, a_2, \ldots, a_n)$. This completes the proof.

We have already mentioned the early milestones in the development of commutative ring theory. Hilbert had two of them, the Hilbert basis theorem and the Nullstellensatz. Emmy Noether provided a third with her groundbreaking paper of 1921. But, without doubt, the next significant milestone was Krull's hugely influential *principal ideal theorem* of 1928, which even today seems almost everywhere present in the study of Noetherian rings. Just to make the point, his theorem is almost always used as the basis for a proof of Theorem 14.5. There are two forms for this theorem; we now state, without proof, the more general form .

Theorem 14.8 (Krull's principal ideal theorem). *Let R be a Noetherian ring. Let $I = (a_1, \ldots, a_k)$ be a proper ideal of R generated by k elements. Let P be a prime ideal of R that is minimal over I. Then $ht(P) \leq k$.*

We can now use Theorem 14.4 to give one final result on the relationship between the dimension of a ring and the dimensions of its polynomial rings. We couldn't ask for a nicer result with which to end this section.

Theorem 14.9. *Let R be a Noetherian ring. Then*

$$\dim R[x_1, \ldots, x_n] = n + \dim R.$$

Proof

First, we observe that the general fact that $\dim R[x_1, \ldots, x_n] \geq n + \dim R$ follows immediately from the lower bound in Theorem 14.2.

Next, we remark that, by Theorem 13.1, $R[x_1, \ldots, x_n]$ is Noetherian, and also, then, by Problem 13.14, for any multiplicative system T, so too is the localization $R[x_1, \ldots, x_n]_T$.

Now, we can begin the proof. Let Q be a maximal ideal in the ring $R[x_1, \ldots, x_n]$, and let $P = Q \cap R$. Consider the multiplicative system $R \setminus P$. By Theorem 6.1, there is a one-to-one correspondence between prime ideals in the localization $R[x_1, \ldots, x_n]_{R \setminus P}$ and prime ideals in $R[x_1, \ldots, x_n]$ disjoint from $R \setminus P$. But, since $Q \cap R = P$, any chain of prime ideals descending from Q will be disjoint from $R \setminus P$; therefore, by virtue of this one-to-one correpondence,

$$ht(Q) = ht\big(Q(R[x_1, \ldots, x_n])_{R \setminus P}\big).$$

Thus we can, without loss, assume that R is a local ring, and that P is its unique prime ideal (see Definition 6.8 and Example 14 in Chapter 6).

We will also need to show that $\mathrm{ht}(P[x_1, \ldots, x_n]) = \mathrm{ht}(P)$. It is for this purpose that we need Krull's principal ideal theorem. Let $\mathrm{ht}(P) = k$.

Since, by Problem 11.10, any chain

$$P = P_0 \supset P_1 \supset P_2 \supset \cdots \supset P_k$$

can immediately be "lifted" to a chain

$$P[x_1, \ldots, x_n] = P_0[x_1, \ldots, x_n]$$

$$\supset P_1[x_1, \ldots, x_n] \supset P_2[x_1, \ldots, x_n] \supset \cdots \supset P_k[x_1, \ldots, x_n],$$

we conclude that $\mathrm{ht}(P[x_1, \ldots, x_n]) \geq \mathrm{ht}(P)$.

Now, by Theorem 14.7, there is an ideal $I = (a_1, \ldots, a_k)$ of R such that P is minimal over I. So, $P[x_1, \ldots, x_n]$ is minimal over $I[x_1, \ldots, x_n]$. But, in the ring $R[x_1, \ldots, x_n]$, the ideal $I[x_1, \ldots, x_n]$ is also generated by the k elements $a_1, \ldots, a_k$. Therefore, by Theorem 14.8 — that is, by Krull's principal ideal theorem — we conclude that $\mathrm{ht}(P[x_1, \ldots, x_n]) \leq k$. Thus, $\mathrm{ht}(P[x_1, \ldots, x_n]) \leq \mathrm{ht}(P)$, and we have shown that

$$\mathrm{ht}(P[x_1, \ldots, x_n]) = \mathrm{ht}(P),$$

as desired.

Finally, then, we can apply Theorem 14.4 — using the fact that R is a local ring with unique maximal ideal P, so that $\mathrm{ht}(P) = \dim R$ — to get

$$\mathrm{ht}(Q) \leq n + \mathrm{ht}(P[x_1, \ldots, x_n])$$

$$= n + \mathrm{ht}(P)$$

$$= n + \dim R.$$

Since Q is an arbitrary maximal ideal in $R[x_1, \ldots, x_n]$, it follows immediately that $\dim R[x_1, \ldots, x_n] \leq n + \dim R$. This completes the proof of the theorem.

The Dimension of Power Series Rings

We turn now to power series and explore the relationship between the dimension of a ring and the dimension of its power series ring. Just as we did for polynomial rings, for a ring R and an ideal I of R, we define $I[[x]]$ to be the set of all power series whose coefficients are in I. Here, too, $I[[x]]$ is the kernel of the natural homomorphism from $R[[x]]$

to $(R/I)[[x]]$ — see Problem 12.9 — and so, in particular, an ideal P is a prime ideal of R if and only if $P[[x]]$ is a prime ideal of $R[[x]]$.

So, in *both* the polynomial ring case and the power series ring case, for a prime P of a ring R, the heights of the prime ideals $P[x]$ and $P[[x]]$ are at least as great as the height of the prime P, simply because any chain of primes descending from P can be "lifted" to chains of the same length descending from $P[x]$ and from $P[[x]]$. One immediate consequence of this is that

$$\text{if } \dim R = \infty, \text{ then } \dim R[x] = \infty \text{ and } \dim R[[x]] = \infty.$$

When the dimension of R is finite, Theorem 14.2 showed that there are strict limitations on the dimension of $R[x]$ in terms of the dimension of R. And, we know that for some classes of rings, such as Noetherian rings, the lower bound is even achievable and we have

$$\dim R[x] = \dim R + 1.$$

However, for power series, the story is very different. It is even possible to have a ring R such that $\dim R = 0$ and yet $\dim R[[x]] = \infty$.

The source of the dramatic difference between the polynomial situation and the power series situation is that in a power series ring $R[[x]]$, the ideal $IR[[x]]$ *generated* by I, while still clearly contained in the ideal $I[[x]]$, may not actually be equal to $I[[x]]$. Here is an example of this unfortunate phenomenon.

Example 4

Let $R = \mathbf{Q} \times \mathbf{Q} \times \mathbf{Q} \times \cdots$ be the ring which is a countably infinite product of the rationals $\mathbf{Q}$. Let I be the ideal of R generated by the countably many elements

$$(1, 0, 0, 0, 0, \ldots), (0, 1, 0, 0, 0, \ldots), (0, 0, 1, 0, 0, \ldots), \ldots .$$

Note, first, that if $a \in I$, then the "vector" a is nonzero in only finitely many places, because a, being a linear combination of a finite number of generators, can be nonzero only where any of these generators are nonzero. Thus we can describe the ideal I by saying it is the set whose elements have *finite support*. So, for example, the element $(1, 0, 1, 0, 1, 0, 1, 0, \ldots)$, where 1 and 0 alternate forever, is not an element of I.

For exactly the same reason, then, if $f \in IR[[x]]$, then f has finite support. But, now, here is an example of a power series in $I[[x]]$ that clearly does not have finite support:

$$g = (1, 0, 0, 0, 0, \ldots) + (0, 1, 0, 0, 0, \ldots)x + (0, 0, 1, 0, 0, \ldots)x^2 + \cdots.$$

Hence, $g \in I[[x]] \setminus IR[[x]]$.

The ring R in Example 4 is an especially nice example of a *von Neumann regular* ring. You saw another example of a von Neumann regular ring in Problem 13.10. A ring is said to be **von Neumann regular** if for each element $a \in R$, there is an element $b \in R$ such that $a = a^2b$. For example, in the ring R above, for an element $a = (1, 0, 2, 0, 3, 0, 4, 0, \ldots)$, the element b would be $(1, 0, \frac{1}{2}, 0, \frac{1}{3}, 0, \frac{1}{4}, 0, \ldots)$. The idea is that b, while not technically an inverse of a, still acts like an inverse where it counts, namely, on the support of a.

Now, without going into any of the lengthy details, we can say a bit more about what happens when $I[[x]]$ and $IR[[x]]$ differ. For a ring R, and an ideal I of R,

$$\text{if } I[[x]] \not\subseteq \sqrt{IR[[x]]}, \text{ then } \dim R[[x]] = \infty.$$

For example, for the power series g from Example 4, $g \in I[[x]]$, but since g does not have finite support, $g \notin \sqrt{IR[[x]]}$. Thus $\dim R[[x]] = \infty$. In Problem 14.14 you are asked to show, as you may have already guessed, that $\dim R = 0$. Problem 14.15 provides another example of this rather astonishing phenomenon.

Let us now turn away from such pathology and return to power series rings that exhibit better behavior.

Example 5

Let K be a field. You are asked to show in Problem 14.16 that the only nonzero ideals of $K[[x]]$ are $(x), (x^2), (x^3), \ldots$; and so, (x) is the unique maximal ideal in $K[[x]]$. Thus, $\dim K[[x]] = 1$, and we have

$$\dim K[[x]] = \dim K + 1.$$

That (x) was the only maximal ideal in $K[[x]]$ should have come as no surprise; after all, in Theorem 12.4 we saw that, for a ring R, the maximal ideals of $R[[x]]$ are just all ideals of the form

$$M + (x),$$

where M is a maximal ideal of R. And, in Problem 12.8, we also saw that if P is a prime ideal in R, then

$$P + (x),$$

is a prime ideal in $R[[x]]$.

We conclude this chapter on dimension with a wonderfully reassuring result which says that, over Noetherian rings, power series rings behave just the way we hope. Once again, and because we are in the context of the dimension of Noetherian rings, we must rely — quite fittingly — on Krull's principal ideal theorem.

Theorem 14.10. *Let R be a Noetherian ring. Then*

$$\dim R[[x]] = \dim R + 1.$$

Proof

Let $\dim R = n$. Then let

$$P = P_0 \supset P_1 \supset P_2 \supset \cdots \supset P_n$$

be a chain of prime ideals of R of length n. As usual, this chain can "lifted" to a chain of the same length in $R[[x]]$, but, we can also add one more prime ideal at the top, and get

$$P + (x) \supset P[[x]] = P_0[[x]] \supset P_1[[x]] \supset P_2[[x]] \supset \cdots \supset P_n[[x]],$$

which is a chain of prime ideals of $R[[x]]$ of length $n + 1$. Thus,

$$\dim R[[x]] \geq \dim R + 1.$$

(The argument so far, of course, has nothing to do with the ring R being Noetherian, and so, the lower bound in Theorem 14.2 holds in general for $R[[x]]$ as well.)

Now, by Theorem 12.4, let $M + (x)$ be any maximal ideal of $R[[x]]$. Since $\mathrm{ht}(M) \leq n$, by Theorem 14.7, there are elements $a_1, \ldots, a_n$ in R such that M is minimal over $(a_1, \ldots, a_n)$. But, then, $M + (x)$ is minimal over $(a_1, \ldots, a_n, x)$. Therefore, by Krull's principal ideal theorem, Theorem 14.8, $\mathrm{ht}(M + (x)) \leq n + 1$. Since $M + (x)$ was an arbitrary maximal ideal of $R[[x]]$, it follows that $\dim R[[x]] \leq n + 1$. Hence,

$$\dim R[[x]] = \dim R + 1,$$

as desired. This completes the proof.

Problems

14.1   Let R be an integral domain such that dim $R = 0$. Prove that R is a field.

14.2   Let R be a principal ideal domain that is not a field. Prove that
       dim $R = 1$.

14.3   Let R be a ring, and let I be an ideal of $R[x]$. Prove that $IR[x] = I[x]$.

14.4   Let R be a ring, and let T be a multiplicative system of R. Verify that

$$(R[x])_T \cong R_T[x].$$

14.5   Let R be a ring such that dim $R = \infty$. Prove that dim $R[x] = \infty$.

14.6   Prove Theorem 14.2.

14.7   Let I be an ideal. Verify that $\sqrt{I}$ is an ideal directly, using the definition
       of ideal.

14.8   Prove that the radical of an ideal I is the intersection of all the prime
       ideals containing I. (Hint: use Zorn's lemma.)

14.9   Let I be an ideal of a ring R. Prove that $\sqrt{I}/I$ is the nilradical of the
       ring R/I.

14.10 Let R be a ring. The **Jacobson radical** of R — written rad(R) — is the
       intersection of all the maximal ideals of R. Prove that if $a \in$ rad(R),
       then $1 + a$ is a unit in R.

14.11 (*Nakayama's lemma*) Let R be a ring and let $J = $ rad(R) be the Jacobson
       radical of R. Let I be a finitely generated ideal of R. Prove that if $JI = I$,
       then $I = (0)$. (Hint: Let I be generated by a *minimal* set of generators,
       and reach a contradiction.)

14.12 Verify in the proof of Theorem 14.6 that $\sqrt{I} = \sqrt{(I, a)} \cap \sqrt{(I, b)}$.

14.13 Let $Q_1, \ldots, Q_k$ be prime ideals of a ring R, and let I be an ideal of R.
       Prove that if $I \subseteq Q_1 \cup \cdots \cup Q_k$, then I is contained in Q_i for some i.

14.14 Let R be the von Neumann regular ring of Example 4. Prove that
       dim $R = 0$.

14.15 Let R be the ring from Problem 12.12. Prove that dim $R = 0$. In fact, R is a local ring with a unique *prime* ideal. Show that $\dim R[[x]] = \infty$ by showing that for this prime ideal, M, $M[[x]] \nsubseteq \sqrt{MR[[x]]}$.

14.16 Let K be a field. Prove that the only nonzero ideals of $K[[x]]$ are $(x), (x^2), (x^3), \ldots$. Thus (x) is the unique maximal ideal in $K[[x]]$, and $K[[x]]$ is a principal ideal domain. $K[[x]]$ is an example of what is called a *rank one discrete valuation ring*.

15

Gröbner Bases

Introduction

We began our journey in ring theory with the story of the Hilbert basis theorem (Theorem 13.1). It is the nonconstructive and existential nature of this famous theorem that caused all of the controversy at the time ("Das ist nicht Mathematik. Das ist Theologie."). In the years that followed mathematicians became quite comfortable with the idea of being able to prove that various things existed without worrying too much about whether or not we could ever actually get our hands on them in any practical or constructive way.

Let's be specific about what the issue is here. For example, we know by the Hilbert basis theorem that if K is a field and I is any ideal in the polynomial ring $K[x_1, x_2, \ldots, x_n]$ over K, then I can be generated by a finite set of polynomials $f_1, f_2, \ldots, f_k$; that is, $I = (f_1, f_2, \ldots, f_k)$. But, and this is the main point, we have no practical way at all of finding these polyomials. Or, suppose $f \in K[x_1, x_2, \ldots, x_n]$ is a polynomial in this ring. How do we determine whether or not $f \in I$? These are straightforward questions and ones we should be able to answer.

And even if we have previously somehow managed to find generators for the ideal I and already know that $f \in I$, how do we then go about expressing f in terms of those generators? That is, how do we find polynomials $h_1, h_2, \ldots, h_k$ such that $f = h_1 f_1 + h_2 f_2 + \cdots + h_k f_k$? There are other questions we really should be able to answer satisfactorily. For example, let J also be an ideal in $K[x_1, x_2, \ldots, x_n]$ and suppose $J = (g_1, g_2, \ldots, g_j)$. Then we surely would like to be able to determine computationally whether $I = J$, or whether one of these ideals is contained in the other, or perhaps even construct the ideal $I \cap J$ in some algorithmic fashion.

As we near the end of our journey in ring theory, it is highly appropriate in this chapter that we try to come to grips with practical questions such as these. Such questions do indeed very much have a nineteenth-century computational feel to them, and the highly computational methods that have been developed within the past several decades to deal with such questions as these bring us back full circle to the very beginnings of our subject.

The key to being able to do the sort of computations needed to answers questions of this kind is what is now called a *Gröbner basis*, which is nothing more than a very special set of generators for an ideal *I*. Without worrying for the moment about exactly what a Gröbner basis is, the idea is simply that once you are able to represent ideals in terms of such a set of polynomials — or, more properly, in terms of a *reduced* Gröbner basis — then it will be possible to proceed algorithmically to deal with each of the questions we have raised, and more.

Example 1

Let **Q** be the field of rationals, and let

$$f = xy^2 - x + y \quad \text{and} \quad g = x^2y - y$$

be two polynomials in $\mathbf{Q}[x, y]$. Let $I = (f, g)$ be the ideal of $\mathbf{Q}[x, y]$ generated by f and g. Next, let

$$h = x^5 - x^4y - 2x^3y^2 + 2x^2y^3 - 2xy^4 - y^5 + 2x^3 - 3x^2y + 4xy^2$$
$$+ y^3 - x - y.$$

Is $h \in I$?

As straightforward as this question is, it is clear that it may not be one that is easy to answer in practice. We will get to the details later, but the first step is to produce a reduced Gröbner basis for the ideal *I*, which in this case turns out to be $\{g_1, g_2, g_3\}$, where

$$g_1 = xy^2 - x + y, g_2 = x^2 - xy - y^2, \text{ and } g_3 = x + y^3 - 2y.$$

The next step — again, the exact details will come later — is to express *h* in terms of this basis, which as it turns out can be done (otherwise, the algorithm being used will automatically let us know that $h \notin I$) as follows:

$$h = (3 - x^2 - y^2)g_1 + (x^3 + y^3)g_2 + (x^2 + 2)g_3,$$

which means that $h \in I$. Moreover, if we choose, we can backtrack through the calculation and even express *h* in terms of the original generators of *I* and get

$$h = (-x^4 + x^3y - xy^3 - 2x^2 + 2xy - y^2 + 1) f$$
$$+ (x^3y - x^2y^2 + y^4 + x^2 - 2y^2 + 2) g.$$

As you can imagine, it would be extremely convenient to be able to use computers to do some, or even all, of the work involved in dealing with problems such as the one posed in Example 1. Fortunately, there are now computer algebra systems — Mathematica, Macsyma, Maple, Macauley, CoCoA, for example — that do just that. In fact, it is no accident that the rapid emergence of computational commutative algebra as a field over the past several decades has coincided with the similarly rapid growth in modern computing. Gröbner bases are so useful these days largely because we have the computing power to be able to handle the massive calculations that are involved in all but the simplest problems.

However, the ideas behind Gröbner bases go as far back as 1900 and Paul Gordan, who used them to give his own constuctive proof of the Hilbert basis theorem. But Gröbner bases themselves were introduced in the 1960s by Bruno Buchberger (and independently by Hironaka Heisuke) in his thesis under the direction of his advisor, Wolfgang Gröbner, who had been using similar ideas for quite some time. Appropriately, it was Buchberger who named his bases "Gröbner bases" to honor his thesis advisor.

Ordering Polynomials

As you might guess, given that the ideas behind Gröbner bases have been around for a long time, the fundamental ideas are fairly natural. For example, if we want to see whether $4,294,967,297 \in (641)$ in the ring of integers $\mathbf{Z}$, we simply divide 641 into $4,294,967,297$ and look at the remainder; similarly, if we want to know whether the polynomial $x^4 + 1 \in (x + 1)$ in the ring $\mathbf{C}[x]$ of polynomials over the complex numbers $\mathbf{C}$, we simply divide $x + 1$ into $x^4 + 1$ and look at the remainder.

So, in Example 1 above, in order to determine whether $h \in I$, we would like to be able to "divide h by both of the two polynomials f and g and look at the remainder." Figuring out how to do this both carefully and in a sufficiently general way will occupy us for the remainder of this chapter. Let us begin by thinking somewhat carefully about how we divide $x + 1$ into $x^4 + 1$. We start by dividing x into x^4 (that is, we divide the highest power term of $x + 1$ into the highest power term of $x^4 + 1$). This gives us x^3, which we then multiply by $x + 1$ and subtract from $x^4 + 1$. The result is a new polynomial $-x^3 + 1$, necessarily having degree less than that of the original polynomial, and we can repeat the process with this new polynomial, dividing x into $-x^3$ to get $-x^2$, and so on.

Since the degree drops after the subtraction at each stage, it is clear that this process will end in one of two ways: either a final subtraction will yield 0, or else it will yield a polynomial (in this case, it would be a nonzero constant) whose degree is less than that of x, bringing the process to an immediate halt. For example, in the division of $x^4 + 1$ by $x + 1$ the final stage ends up being to divide x into the "new" polynomial $-x - 1$ to get -1, and when this is multiplied by $x + 1$ and subtracted from $-x - 1$ the result is 0. Thus, we know that $x^4 + 1 \in (x + 1)$. Moreover, we can keep track of the individual quotients at each stage, which in this example are x^3, $-x^2$, x, and -1, and then write $x^4 + 1 = (x^3 - x^2 + x - 1)(x + 1)$.

Obviously, what makes this division algorithm work for the ring $\mathbf{C}[x]$ is that the terms of polynomials have a natural "ordering." It makes sense to talk about the *highest power term* in a polynomial. It is also always the case that when you can divide one term into another it is the term of lower order being divided into the other term of equal or higher order. (Note that since we are working over a field it is only the order that matters, not the coefficients of the terms involved: $7x^2$ divides $2x^7$ in this ring.) In particular, you can never divide a term of higher order into a term of lower order. Finally, the division is forced to stop at some point. It is precisely these properties that we used above to divide $x + 1$ into $x^4 + 1$. Thus, what makes our division algorithm work in $\mathbf{C}[x]$ is the "ordering"

$$1 < x < x^2 < x^3 < \cdots < x^n < x^{n+1} < \cdots$$

of the powers of x.

So, the first very thing we need to do for a general polynomial ring $K[x_1, x_2, \ldots, x_n]$ over a field K is to come up with an "ordering" of the **monomials** in the ring, that is, of the polynomials in the ring of the form $x_1^{a_1} x_2^{a_2} \cdots x_n^{a_n}$. So, for example, in a ring $K[x, y, z]$ we need to decide in each case which monomial is to be bigger: x^4 or xy^5z^2, x^2 or y^2 or z^2, x^2y or xyz or yz^2. It turns out there are several different satisfactory orderings that can be used for the monomials of $K[x, y, z]$. Before we discuss these, let's clarify what we mean by an "ordering."

Recall from Definitions 4.1 and 4.2 in Chapter 4 that a *total order* is a partial order $\leq$ on a set S such that, for any $x, y \in S$, either $x \leq y$ or $y \leq x$. (In particular, in Chapter 4 on Zorn's lemma and in Chapter 14 on dimension we have previously been concerned with *chains*, that is, with totally ordered subsets that exist within some larger set.) Total orders are just what we need for ordering monomials in polynomial rings.

Definition 15.1. *A* **monomial order** *on the ring $K[x_1, x_2, \ldots, x_n]$ of polynomials over a field K is a total order on the monomials of $K[x_1, x_2, \ldots, x_n]$*

such that, for any monomial $x_1{}^{a_1} x_2{}^{a_2} \cdots x_n{}^{a_n} \neq 1$,

$$1 < x_1{}^{a_1} x_2{}^{a_2} \cdots x_n{}^{a_n},$$

and such that, whenever

$$x_1{}^{a_1} x_2{}^{a_2} \cdots x_n{}^{a_n} < x_1{}^{b_1} x_2{}^{b_2} \cdots x_n{}^{b_n},$$

then, for any monomial $x_1{}^{c_1} x_2{}^{c_2} \cdots x_n{}^{c_n}$,

$$(x_1{}^{a_1} x_2{}^{a_2} \cdots x_n{}^{a_n}) (x_1{}^{c_1} x_2{}^{c_2} \cdots x_n{}^{c_n})$$
$$< (x_1{}^{b_1} x_2{}^{b_2} \cdots x_n{}^{b_n}) (x_1{}^{c_1} x_2{}^{c_2} \cdots x_n{}^{c_n}).$$

There are three especially useful monomial orders on the polynomial ring $K[x_1, x_2, \ldots, x_n]$ over a field K. We will illustrate each of these using the ring $K[x, y, z]$.

Example 2
The first example of a monomial order is called the **lexicographic order**. We will say that

$$x^{a_1} y^{a_2} z^{a_3} <_{\text{lex}} x^{b_1} y^{b_2} z^{b_3}$$

provided that, working from the left, for the first exponents a_i and b_i that differ, we have $a_i < b_i$.

For instance, $xy^2 z <_{\text{lex}} x^2 yz$ since these monomials differ in the very first exponent where $1 < 2$, whereas $x^3 yz <_{\text{lex}} x^3 yz^7$ since these two monomials agree for the first two exponents, but differ in the third where $1 < 7$. In particular, $z <_{\text{lex}} y <_{\text{lex}} x$ and, in fact,

$$1 <_{\text{lex}} z <_{\text{lex}} z^2 <_{\text{lex}} \cdots <_{\text{lex}} y <_{\text{lex}} yz$$
$$<_{\text{lex}} yz^2 <_{\text{lex}} \cdots <_{\text{lex}} y^2 <_{\text{lex}} \cdots <_{\text{lex}} x <_{\text{lex}} \cdots .$$

Note that we would have a very different lexicographic order had we chosen, for example, to let $x < y < z$ or to let $z < x < y$. It is important always to be aware, for any lexicographic order, of the underlying order of the indeterminates. For indeterminates $x_1, x_2, \ldots, x_n$, that standard order is taken to be $x_1 > x_2 > \cdots > x_n$. When we use letters such as x, y, or z for the indeterminates we choose the alphabetical order $x > y > z$.

Example 3
The second example of a monomial order is called the **degree lexicographic order**. As the name suggests, the degree lexicographic order

pays somewhat more attention to the degree of the monomials than does the lexicographic order — where a monomial of low degree such as $x^2 y$ is bigger than a monomial of high degree such as $y^{641} z^{1729}$. We will say that

$$x^{a_1} y^{a_2} z^{a_3} <_{\text{deglex}} x^{b_1} y^{b_2} z^{b_3}$$

if $a_1 + a_2 + a_3 < b_1 + b_2 + b_3$, or, in case $a_1 + a_2 + a_3 = b_1 + b_2 + b_3$, if $x^{a_1} y^{a_2} z^{a_3} <_{\text{lex}} x^{b_1} y^{b_2} z^{b_3}$.

In other words, the degree lexicographic order first compares the total degree of two monomials to see which monomial is greater, and only if the total degrees are equal then "breaks the tie" by reverting to the lexicographic order. For instance, $x^2 yz <_{\text{deglex}} xy^{641} z^{1729}$ simply because of the total degrees, but $xy^2 z <_{\text{deglex}} x^2 yz$ because, even though the total degrees are the same, the tie is resolved by the lexicographic order.

Example 4

The third example of a monomial order is called the **degree reverse lexicographic order**. We will say that

$$x^{a_1} y^{a_2} z^{a_3} <_{\text{degrevlex}} x^{b_1} y^{b_2} z^{b_3}$$

if $a_1 + a_2 + a_3 < b_1 + b_2 + b_3$, or, in case $a_1 + a_2 + a_3 = b_1 + b_2 + b_3$, if, working from the right, for the first exponents a_i and b_i that differ, we have $a_i > b_i$.

Note that the word "reverse" is doing double duty here since not only does it suggest working from-the-right lexicographically in the tie-breaking mode, but it also tells us to reverse the order on the exponents when we get to them where they differ. So, for example, $x^2 y^2 z^3 <_{\text{degrevlex}} x^4 yz^2$ because even though the total degrees are the same, the tie is resolved *from the right* where 3 is *bigger* than 2.

Since our motivation for ordering the monomials of $K[x_1, x_2, \ldots, x_n]$ in the first place was to be able to imitate the ordinary division algorithm for polynomials in $K[x]$ we should make sure we are on the right track by pausing to see that monomial orders do in fact extend the partial order on monomials determined by divisibility. You are asked to do this explicitly in Problem 15.6 for each of the three examples of monomial orders we have just seen, and then to do this in general in Problem 15.7 by showing that for any monomial order $<$ whenever $x_1^{a_1} x_2^{a_2} \cdots x_n^{a_n} | x_1^{b_1} x_2^{b_2} \cdots x_n^{b_n}$, then $x_1^{a_1} x_2^{a_2} \cdots x_n^{a_n} \leq x_1^{b_1} x_2^{b_2} \cdots x_n^{b_n}$.

This does indeed guarantee that we are on the right track for finding an analogue of the division algorithm in which we can use division by monomials, because, just as before, at each stage division will decrease the "degree" of a polynomial. The other key thing to check at this point is that this division can't go on forever. A property we still need, then, is that any *decreasing* chain of monomials must be finite; that is, a decreasing chain of monomials must terminate in a *least* element. Thus, a property we desire for a monomial order is that it be a *well-ordering*, a property we discussed briefly in Chapter 4 in the context of the Axiom of Choice.

Definition 15.2. *A total order on a set S is a* **well-ordering** *if every nonempty subset of S has a least element.*

For monomial orders we record this key property as a theorem.

Theorem 15.1. *Every monomial order on the ring $K[x_1, x_2, \ldots, x_n]$ of polynomials over a field K is a well-ordering.*

Proof
Let $\leq$ be a monomial order. We must prove that any non-empty set of monomials in $K[x_1, x_2, \ldots, x_n]$ has a least element. Assume, by way of contradiction, that there is a non-empty set T of monomials that has no least element. Let $x_1^{a_1} x_2^{a_2} \cdots x_n^{a_n} \in T$. Since this cannot be a least element in T, we can choose another element from T which is smaller. Let that element be $x_1^{b_1} x_2^{b_2} \cdots x_n^{b_n}$. Now, this element can't be a least element of T either, so we can pick a third element which is smaller still. In this way, then, we construct a *strictly* decreasing chain of monomials

$$x_1^{a_1} x_2^{a_2} \cdots x_n^{a_n} > x_1^{b_1} x_2^{b_2} \cdots x_n^{b_n} > x_1^{c_1} x_2^{c_2} \cdots x_n^{c_n} > \cdots$$

in $K[x_1, x_2, \ldots, x_n]$.

This, in turn gives us a strictly *ascending* chain of ideals generated by these monomials:

$$\left(x_1^{a_1} x_2^{a_2} \cdots x_n^{a_n} \right) \subset \left(x_1^{a_1} x_2^{a_2} \cdots x_n^{a_n}, x_1^{b_1} x_2^{b_2} \cdots x_n^{b_n} \right)$$

$$\subset \left(x_1^{a_1} x_2^{a_2} \cdots x_n^{a_n}, x_1^{b_1} x_2^{b_2} \cdots x_n^{b_n}, x_1^{c_1} x_2^{c_2} \cdots x_n^{c_n} \right) \subset \cdots.$$

But this contradicts the fact that $K[x_1, x_2, \ldots, x_n]$ is Noetherian, which, after all, is the Hilbert basis theorem. (Showing that this chain of ideals is *strictly* ascending is left to Problem 15.8.) This contradiction completes the proof.

The Division Algorithm

With Theorem 15.1 in hand we are now ready to start dividing polynomials in $K[x_1, x_2, \ldots, x_n]$ in a systematic way. But, first, let us take a
moment to see what the goal is. In the study of divisibility in number
theory the most basic theorem of all is the one known as the *division
algorithm* which states the all but obvious fact that:

> Given any two integers a and b with $b > 0$, there exist unique
> integers q and r such that $a = qb + r$ with $0 \leq r < b$.

In other words, we can always divide a positive integer b into any integer
a as many times as we like until the "remainder" r is smaller than b;
moreover, we can at this point divide no more, and both the remainder
r and the quotient q — representing the number of divisions by b —
must be unique. Incidentally, the division algorithm for the integers
can be proved using the fact that the natural total order on the positive
integers is a well-ordering (see Problem 15.10). Also, we should mention
that the division algorithm as stated above easily extends to include the
case where b is a negative integer.

There is a similar *division algorithm* for $K[x]$, the ring of polynomials
over a field K:

> Given any two polynomials f and g with $g \neq 0$, there exist
> unique polynomials q and r in $K[x]$ such that $f = qg + r$ with
> $r = 0$ or $\deg(r) < \deg(g)$.

In fact, the algorithm for producing the polynomials q and r is
completely straightforward and is the one we illustrated earlier in the
chapter when we divided the polynomial $x + 1$ into the polynomial
$x^4 + 1$. You simply divide the highest power term of g into the highest
power term of f to get a quotient q_1, and then $f - q_1 g$ is a polynomial
whose degree is strictly less than that of f, and you can then repeat the
process. You keep repeating the process until the degree drops below
that of the divisor g, at which point you have found the remainder r,
and at which point you can also collect together the various individual
quotients to form the quotient q. It is even quite convenient throughout
this process to adopt the standard long-division format in order to keep
track of everything along the way. Verifying the uniqueness of r and q
is left for Problem 15.11.

What we are looking for, then, is a *division algorithm* for the ring
$K[x_1, x_2, \ldots, x_n]$ that will be something like:

> Given any polynomial f, and any nonzero polynomials
> $g_1, \ldots, g_k$, there exist polynomials $q_1, \ldots, q_k$ and a polynomial r
> such that $f = q_1 g_1 + \cdots + q_k g_k + r$.

And, since r is supposed to be the remainder after division by the polynomials $g_1, \ldots, g_k$, we need to have a way to express that r is small and that no further division is possible. (In $K[x]$ this was because either $r = 0$ or $\deg(r) < \deg(g)$.) And, we would like to have uniqueness for r and uniqueness for the quotients $q_1, \ldots, q_k$. And, of course, we want to have an actual algorithm that does all this. And, we would hope to be able to use this division algorithm to answer one of the main questions raised at the very beginning of this chapter, since then we would certainly expect to be able to say that f is in the ideal generated by the polynomials $g_1, \ldots, g_k$ *if and only if* the remainder $r = 0$.

It is time for an example.

Example 5

This example is just a warm-up exercise to get familiar with the idea of dividing polynomials having more than one indeterminate. So, to warm up, we will divide the polynomial $x + 2y + 1$ into the polynomial $x^2y + 3xy - 4y^2$ in the ring $\mathbf{Q}[x, y]$. We are using the lexicographic order where $x > y$. You should note that each of these polynomials has already been written in descending order. In particular, this enables us always to be dividing by the "leading term" in our divisor $x + 2y + 1$ which in this case is x.

So, we begin by dividing x into the leading term x^2y of $x^2y + 3xy - 4y^2$, which gives us xy. We then multiply xy by $x + 2y + 1$ and subtract the product from $x^2y + 3xy - 4y^2$. The result is a new polynomial $-2xy^2 + 2xy - 4y^2$. We now repeat the process, again dividing by x to get $-2y^2$, and then multiplying and subtracting to get another polynomial $2xy + 4y^3 - 2y^2$. Dividing once more by x yields $2y$ for the quotient, and then multiplying and subtracting yields the polynomial $4y^3 - 6y^2 - 2y$. At this point, we can no longer divide by x since under the lexicographic order x is greater than $4y^3$, the leading term of this last polynomial. Thus, $4y^3 - 6y^2 - 2y$ is the remainer. We can now collect the quotients, and write

$$x^2y + 3xy - 4y^2 = (xy - 2y^2 + 2y)(x + 2y + 1) + (4y^3 - 6y^2 - 2y).$$

This division process should have a completely familiar feel to it, and can be made even easier to carry out by using standard long-division notation.

Now might be as good a time as any to introduce some terminology that will be used repeatedly throughout this chapter. The **leading term** of a nonzero polynomial f in $K[x_1, x_2, \ldots, x_n]$ is the term appearing in f whose monomial is greatest in the monomial order on $K[x_1, x_2, \ldots, x_n]$.

This terminology reflects the fact that we almost always choose to write polynomials in descending order. The **leading monomial** of f is simply the monomial from the leading term, and the **leading coefficient** of f is the coefficient of the leading term.

Example 6

Let us now divide the polynomial $x^2y + x^2 + xy^2$ by *both* of the polynomials $x^2 + y$ and $xy - 1$. Again, we will use the lexicographic order where $x > y$. So, each of these polynomials has been written in descending order. You might want to try using standard long-division notation to keep track of everything as we go along.

We begin by dividing x^2 — the leading term of $x^2 + y$ — into the leading term x^2y of $x^2y + x^2 + xy^2$, which gives us y. The next step is to multiply this intermediate quotient y by $x^2 + y$ and subtract the result from $x^2y + x^2 + xy^2$. This gives us a new polynomial $x^2 + xy^2 - y^2$. We can again divide the leading term x^2 of $x^2 + y$ into the leading term x^2 of this new polynomial, and this gives us 1, which we record as an intermediate quotient. We multiply 1 by $x^2 + y$, and subtract the result from $x^2 + xy^2 - y^2$. This gives us a new polynomial $xy^2 - y^2 - y$.

At this point, x^2 no longer divides the leading term of this new polynomial, so we switch to our second divisor $xy - 1$. We continue by dividing the leading term xy of this divisor into the leading term xy^2 of $xy^2 - y^2 - y$, and this gives us y. Of course, this intermediate quotient y is associated with the polynomial $xy - 1$, which needs to be remembered for when we collect quotients at the end. We multiply y by $xy - 1$ and subtract the result from $xy^2 - y^2 - y$. This gives us another new polynomial $-y^2$. This polynomial is now our remainder because neither leading term, x^2 or xy, divides any term in this polynomial, so we have no choice but to stop. We can now collect the quotients — being careful to place them with the appropriate divisor — and write

$$x^2y + x^2 + xy^2 = (y + 1)(x^2 + y) + y(xy - 1) + (-y^2).$$

So, as Example 6 quite clearly shows, there is indeed an algorithm for dividing any polynomial f in $K[x_1, x_2, \ldots, x_n]$ by multiple nonzero polynomials $g_1, \ldots, g_k$, and this algorithm will produce a remainder that is "small" in the sense that no further division is possible of any term in the remainder by any *leading term* among the polynomials $g_1, \ldots, g_k$. We offer one more example to make sure that all the details of this important algorithm are clear.

Example 7

Let us repeat Example 6 and divide the polynomial $x^2y + x^2 + xy^2$ by both of the polynomials $x^2 + y$ and $xy - 1$. This time, however, we will begin by dividing by xy the leading term of $xy - 1$. We are still using the lexicographic order where $x > y$.

So, we begin by dividing xy into $x^2 y$ which gives us x. The next step is to multiply this intermediate quotient x by $xy - 1$, and subtract the result from $x^2 y + x^2 + xy^2$. This gives us a new polynomial $x^2 + x + xy^2$.

At this point xy no longer divides the leading term of this new polynomial, so we switch to our second divisor $x^2 + y$. We can then continue by dividing the leading term x^2 of this divisor into the leading term x^2 of $x^2 + x + xy^2$, and this gives us 1. We multiply 1 by $x^2 + y$ and subtract the result from $x^2 + x + xy^2$. This gives us another new polynomial $x + xy^2 - y$.

At first glance, you might think that this polynomial is our remainder because neither leading term, xy or x^2, divides x, the leading term of $x + xy^2 - y$. But, the leading term xy divides the second term, xy^2. So, we can continue dividing after all. Furthermore, we know that x is destined to be in the final remainder since nothing we do from here on out can involve x.

Thus, we continue by dividing xy into xy^2, which gives us y. We then multiply y by $xy - 1$ and subtract the result from $x + xy^2 - y$. This gives us our next new polynomial x. Now, we really are done because neither leading term, xy or x^2, divides *any* term in this last polynomial. Hence, this last polynomial x is our remainder. We can once again collect the quotients, and write

$$x^2 y + x^2 + xy^2 = (x + y)(xy - 1) + 1 \cdot (x^2 + y) + x.$$

At this stage we pause and summarize the state of things by giving the **division algorithm** for the ring $K[x_1, x_2, \ldots, x_n]$ — with a given monomial order — of polynomials over a field K:

Given any polynomial f, and any nonzero polynomials $g_1, \ldots, g_k$, there exist polynomials $q_1, \ldots, q_k$ and a polynomial r such that

$$f = q_1 g_1 + \cdots + q_k g_k + r$$

and such that either $r = 0$ or else no leading term among the polynomials $g_1, \ldots, g_k$ divides *any* term in the polynomial r.

This division algorithm for $K[x_1, x_2, \ldots, x_n]$ accomplishes most of what we hoped for, since it allows us to divide a polynomial by multiple divisors until the remainder is sufficiently small that further division is impossible. However, unlike the corresponding division algorithms for $\mathbf{Z}$ and $K[x]$, the division algorithm for $K[x_1, x_2, \ldots, x_n]$ does not produce unique remainders and quotients. You may have noticed the occurrence of this extremely unfortunate phenomenon in Examples 6 and 7.

The division algorithm for $K[x_1, x_2, \ldots, x_n]$ has another surprise in store for us, as we see in the next example.

Example 8

Let us now divide the polynomial $y^3 - xy$ by the polynomials $y - 1$ and $xy - 1$, once again using the lexicographic order where $x > y$. We are hoping to resolve the question: is $y^3 - xy \in (y - 1, xy - 1)$?

Writing $y^3 - xy$ as $-xy + y^3$, so that the terms are in descending order, we begin by dividing by y, the leading term of $y - 1$. This gives us $-x$, and after multiplying and subtracting we get a new polynomial $-x + y^3$. Since y does not divide the leading term of this new polynomial, we divide y into y^3 instead. This gives us y^2. Subsequent steps, still dividing by y, give us two more intermediate quotients y and 1, as well as a last new polynomial, $-x + 1$. This polynomial is the remainder since neither y nor xy divides any term in $-x + 1$.

Now, since this remainder is nonzero, we fully expect to be able to conclude that $y^3 - xy \notin (y - 1, xy - 1)$. But this happens to be wrong, as we will soon see when we do the division once again, this time beginning with division by xy, the leading polynomial of the other divisor $xy - 1$. This gives us -1, and after multiplying by -1 and subtracting we get a new polynomial $y^3 - 1$.

At this point xy no longer divides any term of this new polynomial, so we switch to our second divisor $y - 1$. We could of course continue step-by-step with the algorithm, but we immediately notice that, since $y - 1$ divides $y^3 - 1$, the remaining intermediate quotients will just be the individual terms of $y^2 + y + 1$, and therefore that the eventual remainder will be 0. So, $y^3 - xy \in (y - 1, xy - 1)$, after all, and we can write

$$y^3 - xy = (y^2 + y + 1)(y - 1) + (-1)(xy - 1).$$

This example is especially disappointing to us because we were counting on the division algorithm for $K[x_1, x_2, \ldots, x_n]$ to give us a way to decide the question of whether or not a given polynomial f is an element of an ideal I, which is one of the most basic questions we set out trying to answer at the beginning of this chapter. We had hoped to be able to say that $f \in I$ *if and only if* division of f by the generators of I produces a remainder of 0.

Clearly, we need something more to make this all work. Fortunately, help is at hand.

Gröbner Bases

The difficulties we have encountered with the division algorithm for $K[x_1, x_2, \ldots, x_n]$ — not producing a unique remainder, and also not even always producing a remainder 0 in the case when a polynomial is an element of the ideal generated by the divisors — actually have less to do with any inherent flaw in the algorithm, and more to do with the divisors themselves. In order to avoid these difficulties, it turns out that all we have to do is suitably restrict the sets of polynomials that we use as divisors. The key to this is an idea that, with hindsight, now seems quite natural, yet is one that did not emerge until the mid-1960s: the notion of a *Gröbner basis*.

There are several equivalent ways in which to define a Gröbner basis, and we shall choose one which most closely fits our goals in this chapter. Keep in mind that, for an ideal I, a Gröbner basis will simply be a particularly felicitous set of polynomials in I which can be used effectively in applying the division algorithm in various situations.

Definition 15.3. *Let $K[x_1, x_2, \ldots, x_n]$ — with a monomial order — be the ring of polynomials in n indeterminates over a field K. Let I be an ideal in this ring. Then a **Gröbner basis** for the ideal I is a set of polynomials $\{g_1, \ldots, g_k\}$ in I such that, for any nonzero polynomial $f \in I$, then, for some i, the leading term of g_i divides the leading term of f.*

Note how this definition immediately addresses one of the main difficulties we have encountered — namely, that a polynomial f can be an element of an ideal, and yet no term in f is divisible by any of the leading terms of the generators of that ideal. There are, however, several things that should concern us with regard to this definition. How do we know that a given ideal has a Gröbner basis? Even if one exists in principle, how would we go about finding a Gröbner basis for a given ideal? Is there some systematic way to test a given set of polynomials to determine whether or not they form a Gröbner basis? Or even: do the polynomials of a Gröbner basis for an ideal I always generate the ideal I, as the word "basis" would suggest? We are definitely on the right track, but there is much work to be done.

Let us start with examples. We will soon develop a systematic way of determining whether a set of polynomials is a Gröbner basis or not, but for the time being we will need to rely on ad hoc arguments in the following two examples.

Example 9

Consider again the polynomials $y - 1$ and $xy - 1$ from Example 8, still using the lexicographic order where $x > y$. Does the set $\{y - 1, xy - 1\}$

form a Gröbner basis for the ideal $I = (y - 1, xy - 1)$? Given the out-come of Example 8, we suspect not. In order to show this, we need to come up with a nonzero polynomial $f \in I$ whose leading term is *not* divisible by either y or xy, the leading terms of $y - 1$ and $xy - 1$, respectively. But,

$$x - 1 = xy - 1 - xy + x = (xy - 1) - x(y - 1) \in I,$$

and yet, clearly neither y nor xy divides x. So, as we suspected, $\{y - 1, xy - 1\}$ is not a Gröbner basis for I.

Example 10

Consider the polynomials $x + z$ and $y - z$ in $\mathbf{R}[x, y, z]$ with the lexicographic order where $x > y > z$. Does the set $\{x + z, y - z\}$ form a Gröbner basis for the ideal $I = (x + z, y - z)$? Assume not, by way of contradiction. Then, there is a nonzero polynomial $f \in I$ whose leading term is *not* divisible by either x or y, the leading terms of $x + z$ and $y - z$, respectively. Thus, the leading term of f must in fact look like rz^k, where $r \in \mathbf{R}$. Since this is the leading term of f, no other term in f could have an x or a y either; in other words, $f \in \mathbf{R}[z]$.

Since $f \in I$ we can write

$$f = f_1 \cdot (x + z) + f_2 \cdot (y - z),$$

where $f_1, f_2 \in \mathbf{R}[x, y, z]$. Now, in this expresssion replace every y by a z. Since $f \in \mathbf{R}[z]$, the left-hand side does not change at all. On the right-hand side, the polynomial f_1 changes to a polynomial g_1 and the polynomial $y - z$ changes to $z - z = 0$. Thus, we are left with

$$f = g_1 \cdot (x + z),$$

and so $x + z$ divides f, which is a contradiction (either to the fact that $f \in \mathbf{R}[z]$, or to the assumption that x does not divide the leading term of f). Hence, $\{x + z, y - z\}$ is a Gröbner basis for I.

We could have also reached a contradiction here by treating the polynomial f as a function and noticing that $f(-t, t, t) = 0$ for all $t \in \mathbf{R}$. So, since $f \in \mathbf{R}[z]$, $f(t) = 0$ for all $t \in \mathbf{R}$. Hence, $f = 0$, a contradiction.

In Example 8 we witnessed the deeply disturbing situation where a polynomial f could be an element in an ideal, and yet, when we used the division algorithm to divide f by the generators of that ideal, we did not get 0 for the remainder as we expected. Happily, however, if we restrict ourselves to division by polynomials that form a Gröbner basis, the situation is exactly as we had originally hoped.

Theorem 15.2. *Let $K[x_1, x_2, \ldots, x_n]$ — with a monomial order — be the ring of polynomials in n indeterminates over a field K. Let I be an ideal in this ring, and let $\{g_1, \ldots, g_k\}$ be a Gröbner basis for I. Then, for a polynomial $f \in K[x_1, x_2, \ldots, x_n]$, written as*

$$f = q_1 g_1 + \cdots + q_k g_k + r,$$

where either $r = 0$ or no leading term among the polynomials $g_1, \ldots, g_k$ divides any term in the polynomial r, we have

$$f \in I \text{ if and only if } r = 0.$$

Proof

If $r = 0$ then $f \in (g_1, \ldots, g_k) \subseteq I$, and so $f \in I$. Conversely, assume $f \in I$. But $f - r \in I$, and so we conclude that $r \in I$. Thus, $r = 0$, since otherwise no leading term among the polynomials $g_1, \ldots, g_k$ can divide *any* term in the polynomial r, which contradicts the fact that because $\{g_1, \ldots, g_k\}$ is a Gröbner basis for I, by definition, for some i the leading term of g_i would divide the leading term of r. This completes the proof.

Well, this theorem represents real progress because it finally answers one of the main questions raised at the beginning of this chapter: how do we determine whether or not $f \in I$? The answer is that we can use the division algorithm to divide f by the polynomials in a Gröbner basis for I to see if we get 0 for a remainder, or not.

Note that the hypothesis that $\{g_1, \ldots, g_k\}$ be a Gröbner basis for I was only used in the proof of Theorem 15.2 in one direction. So, $r = 0$ is *always* a sufficient condition for $f \in I$, but with a Gröbner basis it is also a necessary condition. This leads immediately to our next theorem.

Theorem 15.3. *Let $\{g_1, \ldots, g_k\}$ be a Gröbner basis for an ideal I. Then*

$$(g_1, \ldots, g_k) = I.$$

Proof

Clearly, $(g_1, \ldots, g_k) \subseteq I$. Now let $f \in I$. Then, by Theorem 15.2, and using the same notation, $r = 0$, which means that $f \in (g_1, \ldots, g_k)$. Thus, $(g_1, \ldots, g_k) = I$, which completes the proof.

In Examples 6 and 7, and again in Example 8, we saw that the division algorithm for $K[x_1, x_2, \ldots, x_n]$ does not always produce unique remainders. However, if we restrict ourselves to division by polynomials that form a Gröbner basis, that extremely awkward situation can no longer occur.

Theorem 15.4. *Let $K[x_1, x_2, \ldots, x_n]$ — with a monomial order — be the ring of polynomials in n indeterminates over a field K. Let I be an ideal in this ring, and let $\{g_1, \ldots, g_k\}$ be a Gröbner basis for I. Let $f \in K[x_1, x_2, \ldots, x_n]$. Then division of f by the polynomials $g_1, \ldots, g_k$ — using the division algoritm — produces a unique remainder r.*

Proof

Suppose that upon division we have two representations for f, and, hence, two remainders r and s where

$$f = q_1 g_1 + \cdots + q_k g_k + r = p_1 g_1 + \cdots + p_k g_k + s.$$

Recall that, by the division algorithm, unless either $r = 0$ or $s = 0$, no leading term among the polynomials $g_1, \ldots, g_k$ divides *any* term in the polynomial r or in the polynomial s. But $r - s \in I$, so by the definition of a Gröbner basis, if $r - s \neq 0$, then for some i the leading term of g_i divides the leading term of $r - s$, which is impossible since the leading term of g_i divides *none* of the terms in r or s. Therefore, $r - s = 0$, and $r = s$, as claimed. This completes the proof.

We can rephrase Theorem 15.4 without making direct reference to the division algorithm as follows: given $f \in K[x_1, x_2, \ldots, x_n]$, there is a *unique* $r \in K[x_1, x_2, \ldots, x_n]$ such that either $r = 0$ or no leading term among the polynomials $g_1, \ldots, g_k$ divides *any* term in the polynomial r, and such that for some $g \in I$, $f = g + r$. In particular, this means that as long as we have a Gröbner basis for I, we get a unique representation of any polynomial modulo I in terms of its remainder; that is, each coset $f + I$ in the quotient ring R/I has a unique representation $r + I$. We record this unexpected and highly useful benefit of Gröbner bases as a theorem.

Theorem 15.5. *Let $K[x_1, x_2, \ldots, x_n]$ — with a monomial order — be the ring of polynomials in n indeterminates over a field K. Let I be an ideal in this ring, and let $\{g_1, \ldots, g_k\}$ be a Gröbner basis for I. Let $f, g \in K[x_1, x_2, \ldots, x_n]$, and let r and s be their remainders, respectively, upon division by the poly-nomials $g_1, \ldots, g_k$. Then*

$$f + I = g + I \ \text{if and only if} \ r = s.$$

Proof

First, assume that $f + I = g + I$. Then, $f - g \in I$, so $r - s \in I$. If $r - s \neq 0$, then for some i the leading term of g_i divides the leading term of $r - s$. But then, exactly as in the proof of Theorem 15.4, unless either $r = 0$ or $s = 0$, no leading term among the polynomials $g_1, \ldots, g_k$ divides *any* term in the polynomial r or in the polynomial s, so it is impossible for the leading term of some g_i to divide the leading term of $r - s$. Therefore, $r - s = 0$, and $r = s$.

Conversely, if $r = s$, then $f - g \in I$, so $f + I = g + I$. This completes the proof.

Let's look at a specific illustration of the uniqueness of remainders using the only example we currently have of a Gröbner basis.

Example 11

In Example 10 we saw that $\{x + z, y - z\}$ is a Gröbner basis in $\mathbf{R}[x, y, z]$ with the lexicographic order where $x > y > z$. Let's divide the polynomial xyz by the two polynomials in this Gröbner basis to see what happens.

We begin by dividing by x, the leading term of $x + z$. This gives us yz, and after multiplying and subtracting we get a new polynomial $-yz^2$. Since x does not divide any term of this new polynomial, we switch to the other divisor $y - z$, and divide y into $-yz^2$. This gives us $-z^2$, which leads immediately to the remainder $-z^3$. We can now collect the quotients and express the result of this division as

$$xyx = yz(x + z) - z^2(y - z) - z^3.$$

Next, we repeat this division, but in the other order, and this time begin by dividing by y, the leading term of $y - z$. This gives us xz, and, after multiplying and subtracting, we get a new polynomial xz^2. Since y does not divide any term of this new polynomial, we switch to the other divisor $x + z$, and divide x into xz^2. This gives us z^2, which leads immediately to the remainder $-z^3$. We can now collect the quotients and express the result of this division as

$$xyx = xz(x + z) + z^2(y - z) - z^3.$$

Of course, as expected, we do get the same remainder with both calculations, but — and this is the real point of this example — the quotients are not the same. In other words, we do not have uniqueness of quotients, even when using Gröbner bases!

Finding Gröbner Bases

We now turn to the main remaining question: how do we find a Gröbner basis for an ideal I? Or, perhaps we should even first wonder whether every nonzero ideal *has* a Gröbner basis. Fortunately, it does, so the real issue that need concern us is the practical one. But first things first.

Theorem 15.6. *Let $K[x_1, x_2, \ldots, x_n]$ — with a monomial order — be the ring of polynomials in n indeterminates over a field K. Let I be a non-zero ideal in this ring. Then I has a Gröbner basis.*

Proof

Let J be the ideal of $K[x_1, x_2, \ldots, x_n]$ generated by the leading terms of *all* the polynomials in I. Then, by the Hilbert basis theorem, J is finitely generated, so $J = (g_1, \ldots, g_k)$ for some finite set of polynomials $g_1, \ldots, g_k \in K[x_1, x_2, \ldots, x_n]$.

Now, in fact, not only do the polynomials $g_1, \ldots, g_k$ generate the ideal J, but we claim that their *leading terms* also generate the ideal J. Clearly, these leading terms are in J since the ideal J is generated by *all* leading terms of polynomials in I.

Suppose, conversely, that a polynomial $f \in J$; then we can write $f = f_1 g_1 + \cdots + f_k g_k$, so the leading term of f is a linear combination of leading terms of some of the f_is and g_is. Thus, the leading term of f is in the ideal generated by the leading terms of $g_1, \ldots, g_k$. Now, simply repeat this argument on the polynomial f minus its leading term. Inductively, in this way, we see, term by term, that *each* term of f is in the ideal generated by the leading terms of $g_1, \ldots, g_k$; hence, f is also in this ideal, as claimed.

Finally, if $f \in I$, then, by definition of J, the leading term of f can be written as a finite sum of polynomials times the leading terms of the g_is. But then this leading term of f is necessarily divisible by one of the leading terms of of the g_is. That is, the set $\{g_1, \ldots, g_k\}$ forms a Gröbner basis for I, as desired. This completes the proof.

That takes care of existence. The next step is to describe an efficient way to tell if a given set of polynomials in an ideal I is in fact a Gröbner basis for I. For this we need a definition.

Definition 15.4. *Let $K[x_1, x_2, \ldots, x_n]$ — with a monomial order — be the ring of polynomials in n indeterminates over a field K. Let*

$$f = ax_1{}^{a_1} x_2{}^{a_2} \cdots x_n{}^{a_n} + \cdots \in K[x_1, x_2, \ldots, x_n]$$

and

$$g = bx_1{}^{b_1} x_2{}^{b_2} \cdots x_n{}^{b_n} + \cdots \in K[x_1, x_2, \ldots, x_n]$$

have leading terms $ax_1^{a_1}x_2^{a_2}\cdots x_n^{a_n}$ *and* $bx_1^{b_1}x_2^{b_2}\cdots x_n^{b_n}$, *respectively. Let* $c_i = \max(a_i, b_i)$ *for each* i — *so that* $x_1^{c_1}x_2^{c_2}\cdots x_n^{c_n}$ *is the least common multiple of* $x_1^{a_1}x_2^{a_2}\cdots x_n^{a_n}$ *and* $x_1^{b_1}x_2^{b_2}\cdots x_n^{b_n}$. *Then the polynomial*

$$S(f,g) = \frac{x_1^{c_1}x_2^{c_2}\cdots x_n^{c_n}}{ax_1^{a_1}x_2^{a_2}\cdots x_n^{a_n}}\, f - \frac{x_1^{c_1}x_2^{c_2}\cdots x_n^{c_n}}{bx_1^{b_1}x_2^{b_2}\cdots x_n^{b_n}}\, g$$

is called the **S-polynomial** *of f and g.*

Example 12
Consider the polynomials $f = 3x^2y + xy$ and $g = 4x^3 - y^2$ in $\mathbf{R}[x, y]$ with the lexicographic order where $x > y$. Then, the monomial that is the least common multiple for x^2y and x^3 is x^3y. Therefore,

$$S(f,g) = \frac{x^3y}{3x^2y}\, f - \frac{x^3y}{4x^3}\, g = \frac{x}{3}\, f - \frac{y}{4}\, g = x^3y + \frac{x^2y}{3} - x^3y + \frac{y^3}{4}$$

$$= \frac{x^2y}{3} + \frac{y^3}{4}.$$

Now, here is the core of this rather clever idea, since it is quite clear — for instance, from Example 12 — that the point of S-polynomials is that you get cancellation of leading terms. For any two polynomials g_{i_1} and g_{i_2} in a set of polynomials $\{g_1, \ldots, g_k\}$ that we might hope to show is a Gröbner basis for an ideal I, it would certainly be necessary that their S-polynomial $S(g_{i_1}, g_{i_2})$ have a remainder of 0 when divided by the polynomials $g_1, \ldots, g_k$, since this S-polynomial is obviously an element of I. It is extraordinary that such a remarkably obvious necessary condition is also a *sufficient* condition for a set to be a Gröbner basis. This was first proven by Buchberger. His theorem not only gives us a practical way to test sets, but will also soon allow us to construct Gröbner bases algorithmically.

In the next example, we see how we can in some cases express a polynomial in terms of S-polynomials. By doing this we inductively reduce the overall monomial ordering of a polynomial expression within the monomial order. This will be a key step in the proof of Buchberger's theorem.

Example 13
Let a set of polynomials $\{g_1, \ldots, g_j\} \in K[x_1, x_2, \ldots, x_n]$, the ring — with a monomial order — of polynomials in n indeterminates over a field K, be such that all the g_i have *exactly* the same leading monomial $x_1^{b_1}x_2^{b_2}\cdots x_n^{b_n}$. And let $g = b_1g_1 + \cdots + b_jg_j$, where each $b_i \in K$,

be a polynomial such that the leading monomial of g is *less than* $x_1^{b_1} x_2^{b_2} \cdots x_n^{b_n}$ in the monomial order. Our goal in this example is to express g as a linear combination of S-polynomials.

First, letting a_i be the leading coefficient of g_i for each i, we notice that $b_1 a_1 + \cdots + b_j a_j = 0$, simply because of the hypothesis that the leading monomial of g is less than $x_1^{b_1} x_2^{b_2} \cdots x_n^{b_n}$.

Next, we notice that since each g_i has the same leading monomial — which makes it particularly easy to figure out least common multiples — the S-polynomial for any pair g_i and g_k will be just

$$S(g_i, g_k) = \frac{1}{a_i} g_i - \frac{1}{a_k} g_k.$$

This form of the S-polynomial, in turn, suggests the following very fancy bit of telescoping (I urge you to multiply out the terms to better see how nicely the cancelling is working):

$$g = b_1 g_1 + \cdots + b_j g_j$$

$$= b_1 a_1 \left(\frac{1}{a_1} g_1 - \frac{1}{a_2} g_2 \right) + (b_1 a_1 + b_2 a_2) \left(\frac{1}{a_2} g_2 - \frac{1}{a_3} g_3 \right)$$

$$+ (b_1 a_1 + b_2 a_2 + b_3 a_3) \left(\frac{1}{a_3} g_3 - \frac{1}{a_4} g_4 \right)$$

$$+ \cdots$$

$$+ (b_1 a_1 + \cdots + b_{j-1} a_{j-1}) \left(\frac{1}{a_{j-1}} g_{j-1} - \frac{1}{a_j} g_j \right)$$

$$+ (b_1 a_1 + \cdots + b_j a_j) \left(\frac{1}{a_j} g_j \right).$$

But, as we already noticed, the coefficient of this last term is 0, so we have successfully expressed g as a linear combination of the S-polynomials, $S(g_1, g_2), S(g_2, g_3), S(g_3, g_4), \ldots, S(g_{j-1}, g_j)$.

Now, we are ready for Buchberger's theorem, which gives us a practical way to test whether or not a given set of polynomials in an ideal I is in fact a Gröbner basis for I. This theorem tells us that all we have to do is check a finite number of S-polynomials to see if their remainders are 0.

Theorem 15.7. *Let $\{g_1, \ldots, g_k\}$ be a set of polynomials in the ring of polynomials $K[x_1, x_2, \ldots, x_n]$ — with a monomial order — in n indeterminates over a field K, and let $I = (g_1, \ldots, g_k)$ be the ideal generated by this set of polynomials. Then, if it is true that, for all $i \neq j$, whenever the S-polynomial*

$S(g_i, g_j)$ *is divided by the polynomials* $g_1, \ldots, g_k$ *the remainder is 0, then* $\{g_1, \ldots, g_k\}$ *is a Gröbner basis for* I.

Proof

This will be a proof by contradiction, and at the very heart of the contradiction is the fundamental fact that in a well-ordering of monomials every non-empty set of monomials has a *least* element. (Thus, we are using in this proof Fermat's wonderful *method of infinite descent* mentioned in Chapter 9.)

Let $f \in I$. Then, among all the possible ways in which we could represent f in terms of the generating polynomials $g_1, \ldots, g_k$, we shall write

$$f = f_1 g_1 + \cdots + f_k g_k,$$

where the polynomials $f_1, \ldots, f_k$ have been chosen so that the greatest monomial that occurs on the right-hand side of this expression is *least* possible.

Now, of course, if this greatest monomial on the right-hand side is also the leading monomial for f then we are done, since the leading monomial of one of the g_is divides the leading term of f, thus making $\{g_1, \ldots, g_k\}$ a Gröbner basis for I.

Therefore, we now let $x_1^{a_1} x_2^{a_2} \cdots x_n^{a_n}$ be the greatest monomial from the right-hand side (remember, it is the smallest one possible), and assume that the leading monomial of f is less than $x_1^{a_1} x_2^{a_2} \cdots x_n^{a_n}$. In the expression above for f, this greatest monomial $x_1^{a_1} x_2^{a_2} \cdots x_n^{a_n}$ certainly appears in some of the individual products $f_i g_i$, but perhaps not in others. For each i for which $x_1^{a_1} x_2^{a_2} \cdots x_n^{a_n}$ actually appears in $f_i g_i$, we construct a polynomial g by taking the leading term $b_i x_1^{b_{i1}} x_2^{b_{i2}} \cdots x_n^{b_{in}}$ of f_i, and then letting

$$g = b_{i_1} x_1^{b_{i_1 1}} x_2^{b_{i_1 2}} \cdots x_n^{b_{i_1 n}} g_{i_1} + \cdots + b_{i_j} x_1^{b_{i_j 1}} x_2^{b_{i_j 2}} \cdots x_n^{b_{i_j n}} g_{i_j}.$$

Note, before we continue, that all the terms that we used to build g were already present on the right-hand side of our original expression for f; hence, g is also present in this expression.

By design, each of the j products on the right in our expression for g has exactly the same leading monomial $x_1^{a_1} x_2^{a_2} \cdots x_n^{a_n}$ — that's how we chose each i in the first place — and, of course, the leading monomial of g is strictly less than $x_1^{a_1} x_2^{a_2} \cdots x_n^{a_n}$ because this also holds for f, by assumption. So, by Example 13, g can be written as a linear combination of S-polynomials using the set of polynomials

$$\{x_1^{b_{i_1 1}} x_2^{b_{i_1 2}} \cdots x_n^{b_{i_1 n}} g_{i_1}, \ldots, x_1^{b_{i_j 1}} x_2^{b_{i_j 2}} \cdots x_n^{b_{i_j n}} g_{i_j}\}.$$

Now, by hypothesis, we know that when any S-polynomial $S(g_{i_s}, g_{i_t})$ is divided by the polynomials $g_1, \ldots, g_k$ the remainder is 0. It follows easily — this is left as an exercise in Problem 15.20 — that for any two of the polynomials

$$x_1^{b_{i_1 1}} x_2^{b_{i_1 2}} \cdots x_n^{b_{i_1 n}} g_{i_1}, \ldots, x_1^{b_{i_j 1}} x_2^{b_{i_j 2}} \cdots x_n^{b_{i_j n}} g_{i_j},$$

when the S-polynomial

$$S(x_1^{b_{i_s 1}} x_2^{b_{i_s 2}} \cdots x_n^{b_{i_s n}} g_{i_s}, x_1^{b_{i_t 1}} x_2^{b_{i_t 2}} \cdots x_n^{b_{i_t n}} g_{i_t})$$

is divided by the polynomials $g_1, \ldots, g_k$ the remainder is also 0.

In particular, then, each S-polynomial

$$S(x_1^{b_{i_s 1}} x_2^{b_{i_s 2}} \cdots x_n^{b_{i_s n}} g_{i_s}, x_1^{b_{i_t 1}} x_2^{b_{i_t 2}} \cdots x_n^{b_{i_t n}} g_{i_t})$$

can be represented as a linear combination of the polynomials $g_1, \ldots, g_k$, and, of course, its leading monomial is less than $x_1^{a_1} x_2^{a_2} \cdots x_n^{a_n}$. But the leading monomial of g is also less than $x_1^{a_1} x_2^{a_2} \cdots x_n^{a_n}$, since g is now written as a linear combination of these S-polynomials.

Therefore, since g appears in the original expression for f, we have successfully expressed f in terms of $g_1, \ldots, g_k$, where the greatest monomial order is now less than $x_1^{a_1} x_2^{a_2} \cdots x_n^{a_n}$, which by assumption was the *least* such monomial order. This contradiction completes the proof.

Let's take a look at Buchberger's theorem in action.

Example 14

Consider once again the polynomials $y - 1$ and $xy - 1$ from Examples 8 and 9 still using the lexicographic order where $x > y$. We already know that the set $\{y - 1, xy - 1\}$ does not form a Gröbner basis for the ideal $I = (y - 1, xy - 1)$. Let's confirm this using the necessity of the condition in Theorem 15.7. Since there are only two polynomials in this set, there is only one pair whose S-polynomial needs to be checked. We get

$$S(y - 1, xy - 1) = \frac{xy}{y}(y - 1) - \frac{xy}{xy}(xy - 1)$$

$$= xy - x - xy + 1 = -x + 1.$$

Clearly, the S-polynomial $-x + 1$ does not reduce to 0 since neither leading term y nor xy divides either term of $-x + 1$. Therefore, the set $\{y - 1, xy - 1\}$ cannot form a Gröbner basis for I.

Example 15

Consider once again the polynomials $x + z$ and $y - z$ from Example 10, still using the lexicographic order where $x > y > z$. We already know that the set $\{x + z, y - z\}$ forms a Gröbner basis for the ideal $I = (x + z, y - z)$. Let's confirm this using Theorem 15.7. Since there are only two polynomials in this set, there is only one pair whose S-polynomial needs to be checked. We get

$$S(x + z, y - z) = \frac{xy}{x}(x + z) - \frac{xy}{y}(y - z)$$

$$= xy + yz - xy + xz = xz + yz.$$

It is easy to verify that the S-polynomial $xz + yz$ has remainder 0 when divided by $x + z$ and $y - z$ — this is left as an exercise in Problem 15.21. Therefore, by Theorem 15.7, the set $\{x + z, y - z\}$ forms a Gröbner basis for the ideal I.

In Problem 15.22 you are asked to verify that a particular set of three polynomials is a Gröbner basis. In this case there are three S-polynomials to check. In a set with m polynomials, there would be $\frac{m(m-1)}{2}$ S-polynomials to check. This should certainly make it clear why modern computer algebra systems are so helpful in the study of Gröbner bases.

It is also worth mentioning that whether or not a given set of polynomials is a Gröbner basis can depend upon the particular mono-mial order being used. For example, Problem 15.23 contains a set of polynomials that is a Gröbner basis under one monomial order, but is not a Gröbner basis under a different monomial order. This additional flexibility explains why it is useful to have a variety of monomial orders from which to choose.

Well, we are finally ready to tackle the main practical question in this section: how do we find a Gröbner basis for an ideal I? Theorem 15.7 provides a straightforward and simple method. Suppose $I = (f_1, \ldots, f_k)$ is an ideal. Now, the set $\{f_1, \ldots, f_k\}$ may already be a Gröbner basis for I, but, if it isn't, then, by Theorem 15.7, one of the S-polynomials $S(f_i, f_j)$ has a nonzero remainder when divided by $f_1, \ldots, f_k$. So, we can include that remainder in the set of polynomials. The new larger set of poly-nomials still generates I, and obviously now the S-polynomial $S(f_i, f_j)$ has a remainder 0 when divided by this larger set of polynomials. So, this new set now has a chance of being a Gröbner basis for I. But, if not, we can repeat this process, over and over if necessary, adding nonzero remainders to the set. When the process stops, and all the remainders are 0, then we know we have a Gröbner basis for I. This method for finding Gröbner bases is called **Buchberger's algorithm**. Before we verify that this process must always stop, let's look at an example.

Example 16

We will use Buchberger's algorithm to find a a Gröbner basis for the ideal $I = (x^2, xy + y^2)$ in $\mathbf{R}[x, y]$ with the lexicographic order where $x > y$. First, we compute

$$S(x^2, xy + y^2) = \frac{x^2 y}{x^2} x^2 - \frac{x^2 y}{xy}(xy + y^2) = x^2 y - x^2 y - xy^2 = -xy^2.$$

Dividing the S-polynomial $-xy^2$ by $xy + y^2$, we get a quotient $-y$. Multiplying and subtracting produces a remainder y^3, since no further division by either $xy + y^2$ or x^2 is possible. So, we include the remainder y^3 in the original set $\{x^2, xy + y^2\}$ to get a new set of polynomials $\{x^2, xy + y^2, y^3\}$. We now test this new set to see whether it is a Gröbner basis. Of course, we need not check the S-polynomial $S(x^2, xy + y^2)$ again, since we know that this time it will have a remainder 0. The details of checking that both the other S-polynomials $S(x^2, y^3)$ and $S(xy + y^2, y^3)$ have remainder 0 are left as an exercise in Problem 15.22. Thus, we can stop, since we know we have a Gröbner basis at this point.

Why does Buchberger's algorithm stop? Not too surprisingly, it is because $K[x_1, x_2, \ldots, x_n]$ is a Noetherian ring. At any given stage in the process of implementing the algorithm we can associate the current set of polynomials with the ideal generated by all the leading terms of the polynomials in the set. At the next stage, when we have added one remainder to this set, it is clear that the ideal associated with the new set will be a strictly larger ideal simply because the leading term of the remainder we just added could not have been in the previous ideal (see Problem 15.15). Thus, Buchberger's algorithm produces — as a byproduct — an *ascending chain* of ideals, which must terminate since this is all taking place in a Noetherian ring. Therefore, Buchberger's algorithm stops.

For example, in Example 16 the set $\{x^2, xy + y^2\}$ would have the ideal (x^2, xy) associated with it, and the next set $\{x^2, xy + y^2, y^3\}$ would be associated with the ideal (x^2, xy, y^3). This second ideal is a strictly larger ideal because y^3 — being the leading term of the remainder — could not be in the previous ideal by Problem 15.15.

Although we have achieved our primary goal of being able to construct Gröbner bases, there are still a few improvements to be made. Obviously, Buchberger's algorithm tends to produce Gröbner bases that have lots of elements, and there may well be some redundancy. In

particular, if g_i and g_j are two polynomials in a Gröbner basis such
that the leading monomial of g_i divides the leading monomial of g_j,
then g_j is redundant as far as the Gröbner basis goes and can simply
be removed. Clearly, it would be a good idea in practice to remove any
such redundant polynomials. It is also convenient if all the polynomials
in a Gröbner basis have leading terms with coefficient 1, and it is
easy to "normalize" any Gröbner basis in this way simply by dividing
each polynomial in the basis by its leading coefficient. With these
improvements in mind we call a Gröbner basis a **minimal Gröbner
basis** if all the leading coefficients of the polynomials in the basis are 1,
and if in no case does the leading term of one of the polynomials in the
basis divide the leading term of any other polynomial in the basis.

There is one more improvement that we will make. Minimal Gröbner
bases are not unique by any means. Changing the order in which things
are done in Buchberger's algorithm, or when removing redundant polynomials can result in, for a given ideal I, different minimal Gröbner
bases for I. The following definition will achieve uniqueness.

Definition 15.5. *Let $K[x_1, x_2, \ldots, x_n]$ — with a monomial order — be the
ring of polynomials in n indeterminates over a field K. Let I be an ideal in this
ring. Then, a Gröbner basis $\{g_1, \ldots, g_k\}$ for the ideal I is called a **reduced
Gröbner basis** if the leading coefficient of each g_i is 1, and if no term in any
g_j is divisible by the leading term of any of the other g_is.*

There is a completely straightforward procedure for turning a minimal Gröbner basis $\{g_1, \ldots, g_k\}$ into a reduced Gröbner basis. The procedure takes place one polynomial at a time, starting with g_1 and ending
with g_k. At each stage, for a polynomial g_i in the minimal Gröbner
basis, the idea is to "reduce" it by dividing by all the other polynomials
in the basis. Then simply replace g_i by the remainder polynomial that
results from this division. Since this remainder still has the same leading
polynomial as g_i, the new basis will also be a minimal Gröbner basis.
The end result of this procedure is then clearly a reduced Gröbner basis.

Theorem 15.8. *Let $K[x_1, x_2, \ldots, x_n]$ — with a monomial order — be the ring
of polynomials in n indeterminates over a field K. Let I be a nonzero ideal in
this ring. Then I has a unique reduced Gröbner basis.*

Proof
Since we have already seen how it is possible to contruct reduced
Gröbner bases, all that remains to be proved is uniqueness. Suppose that
$A = \{f_1, \ldots, f_j\}$ and $B = \{g_1, \ldots, g_k\}$ are two reduced Gröbner bases for
I. The first claim we make is that the leading terms of the polynomials in

these two sets are exactly the same. (In fact, this would be true even if A and B were merely minimal Gröbner bases.) Without loss of generality, consider the leading term of f_1. Since $f_1 \in I$, the leading term of some polynomial in B, call it g_i, divides the leading term of f_1, but then by exactly the same argument the leading term of some polynomial in A divides the leading term of g_i. This can only be f_1, since otherwise we would have the leading term of one polynomial in A dividing the leading term of another polynomial in A. Since the leading terms of f_1 and g_i divide each other, they are equal. Thus, the leading terms of the polynomials in these two sets match up in a one-to-one fashion, as claimed. (In particular, then, this means that any two minimal Gröbner bases have the same number of elements.)

The second claim is that not only are the leading terms of f_1 and g_i equal, but $f_1 = g_i$. Consider the polynomial $f_1 - g_i$. None of the terms in this polynomial are divisible by any leading term of a polynomial from A, because B has the same leading terms and because the leading terms of f_1 and g_i cancelled each other out. In other words, $f_1 - g_i$ can't be reduced at all by the polynomials in A, but A is a reduced Gröbner basis, so $f_1 - g_i = 0$, as claimed. Therefore, the two reduced Gröbner bases A and B are equal. This completes the proof.

Here is an unexpected payoff from the uniqueness of reduced Gröbner bases. One of the questions raised at the beginning of this chapter was the question of how to determine computationally whether or not $I = J$ for two ideals I and J. In some sense we have been able to answer that question for a long time, because if $I = (f_1, \ldots, f_j)$ and $J = (g_1, \ldots, g_k)$, then all we have to do is check, polynomial by polynomial, that all the fs are in J and all the gs are in I, and we can do this using Theorem 15.2. Now, however, we have a much nicer way to do this. We can simply compute the reduced Gröbner basis for each ideal I and J to see if they are same or not.

We conclude this chapter, and this brief introduction to Gröbner bases, by remarking that we are now ready to answer the question that was posed in Example 1: is the polynomial

$$x^5 - x^4y - 2x^3y^2 + 2x^2y^3 - 2xy^4 - y^5 + 2x^3 - 3x^2y + 4xy^2 + y^3 - x - y$$

in the ideal $I = (xy^2 - x + y, x^2y - y)$? What needs to be done to answer this question is now clear. We construct a reduced Gröbner basis for I and then use the division algorithm to express this polynomial in terms of that basis. If the remainder is 0, the polynomial is in the ideal; otherwise, it isn't. I'll leave the details of this exercise for you to carry out in Problem 15.25.

Problems

Ordering polynomials

15.1 Write the polynomial $3x + y - 2z + 5xy^2z^3 - 7xyz^4 + x^5 - y^3z^2$ in descending order using the lexicographic order, then using the degree lexicographic order, and finally using the degree reverse lexicographic order.

15.2 Lexicographically order the following three monomials in the ring $\mathbf{Z}[a, b, c, \ldots, z]$:

$$abracadabra, \ ab^2rac^2ad^2abr^2a, \ a^5b^3c^3d^2r^2.$$

15.3 Prove that in a ring $K[x, y]$ the degree reverse lexicographic order is the same as the degree lexicographic order. Then show by an example that they are not the same in $K[x, y, z]$.

15.4 Prove that $<_{\text{lex}}$ is a monomial order.

15.5 Prove that $<_{\text{deglex}}$ is a monomial order.

15.6 Prove that the lexicographic order, the degree lexicographic order, and the degree reverse lexicographic order all extend the partial order on monomials in $K[x_1, x_2, \ldots, x_n]$ given by divisibility by explicitly proving for each of the three orders $<_{\text{lex}}$, $<_{\text{deglex}}$, and $<_{\text{degrevlex}}$ that, whenever

$$x_1^{a_1} x_2^{a_2} \cdots x_n^{a_n} \mid x_1^{b_1} x_2^{b_2} \cdots x_n^{b_n},$$

we have

$$x_1^{a_1} x_2^{a_2} \cdots x_n^{a_n} \le x_1^{b_1} x_2^{b_2} \cdots x_n^{b_n}.$$

15.7 Prove that a monomial order always extends the partial order on monomials in $K[x_1, x_2, \ldots, x_n]$ given by divisibility by proving in general that, whenever

$$x_1^{a_1} x_2^{a_2} \cdots x_n^{a_n} \mid x_1^{b_1} x_2^{b_2} \cdots x_n^{b_n},$$

we have

$$x_1^{a_1} x_2^{a_2} \cdots x_n^{a_n} \le x_1^{b_1} x_2^{b_2} \cdots x_n^{b_n}.$$

15.8 Verify in the proof of Theorem 15.1 that the given ascending chain of ideals is a *strictly* ascending chain.

15.9   Prove that a total order on a set S is a well-ordering if and only if
         every decreasing chain of elements of S terminates. (A decreasing
         chain $a_1 \geq a_2 \geq a_3 \geq \cdots$ is said to *terminate* if, for some i,
         $a_i = a_{i+1} = a_{i+2} = \cdots$.)

The division algorithm

15.10   The statement that the natural total order on the non-negative
         integers is a well-ordering is sometimes called the *well-ordering principle*
         (see page 43). Use the well-ordering principle to prove the division
         algorithm for the integers. (Hint: use a and b to construct a set T of
         non-negative integers in such a way that the *least* element of T will be
         the remainder r.)

15.11   Verify that in the division algorithm for $K[x]$ both the quotient q and
         the remainder r are unique.

15.12   Since the polynomial ring $K[x]$ over a field K is a principal ideal
         domain, we know that any ideal is generated by a single polynomial,
         but how do we find that generator? If $I = (f_1, \ldots, f_k)$ then the
         generator is just a *greatest common divisor* of $f_1, \ldots, f_k$. (A polynomial f
         is called a **greatest common divisor** of $f_1, \ldots, f_k$ if f divides each of
         $f_1, \ldots, f_k$ and if, whenever g is any other common divisor of each of
         $f_1, \ldots, f_k$, then g divides f.)
              Find a single generator for the ideal $(x^2 + x - 2, x^3 - x^2 + x - 1)$
         in $\mathbf{Q}[x]$ by using the Euclidean algorithm to find a greatest common
         divisor of $x^2 + x - 2$ and $x^3 - x^2 + x - 1$ (see Problem 10.5). Note that
         once you have the greatest common divisor you can work backward
         to express the greatest common divisor in terms of the original
         polynomials.
              This algorithm can also be used to find the greatest common divisor
         of any number of polynomials since, for example, we can compute
         $\gcd(f_1, f_2, f_3) = \gcd(\gcd(f_1, f_2), f_3)$ always working with only two
         polynomials at a time.

15.13   Use the division algorithm for $K[x, y]$ with the lexicographic order
         where $x > y$ to divide the polynomial $xy^2 - x$ by the polynomials
         $xy + 1$ and $y^2 - 1$. Do the algorithm twice, the first time beginning
         with the divisor $xy + 1$, and then a second time beginning with the
         divisor $y^2 - 1$. Compare the results. In particular, what does this tell us
         about the question: is $xy^2 - x \in (xy + 1, y^2 - 1)$?

15.14   Explain why the division algorithm for $K[x]$ is not very helpful in
         deciding whether $x^2 + x \in (x^3, x^3 - x)$. Nonetheless, it is easy to see

that, in fact, $x^2 + x \in (x^3, x^3 - x)$. (Why?) Express $x^2 + x$ in terms of the two generators x^3 and $x^3 - x$; that is, find polynomials $f, g \in K[x]$ such that $x^2 + x = f \cdot x^3 + g \cdot (x^3 - x)$.

15.15   Let I be an ideal in $K[x_1, x_2, \ldots, x_n]$ that is generated by monomials. Such an ideal is called a **monomial ideal**, and the generating set of monomials may be finite or infinite. Prove that a monomial $x_1^{a_1} x_2^{a_2} \cdots x_n^{a_n}$ in $K[x_1, x_2, \ldots, x_n]$ is an element of the ideal I if and only if $x_1^{a_1} x_2^{a_2} \cdots x_n^{a_n}$ is divisible by one of the generating monomials of I.

Gröbner Bases

15.16   Consider the ideal $I = (xy + 1, yz + 1)$ in $\mathbf{Q}[x, y, z]$. Show that the polynomial $z - x \in I$. For which of the three monomial orders — lexicographic, degree lexicographic, and degree reverse lexicographic — can the set $\{xy + 1, yz + 1\}$ be a Gröbner basis for I?

15.17   Prove that the converse of Theorem 15.2 is also true by showing that the condition that $f \in I$ if and only if division of f by $g_1, \ldots, g_k$ produces a remainder 0 implies that $\{g_1, \ldots, g_k\}$ is a Gröbner basis.

15.18   Prove that the converse of Theorem 15.4 is also true by showing that if division of f by $g_1, \ldots, g_k$ produces a *unique* remainder, then $\{g_1, \ldots, g_k\}$ is a Gröbner basis.

15.19   Prove that if $g_1, \ldots, g_k \in I$ are such that the ideal generated by the leading terms of $g_1, \ldots, g_k$ equals the ideal generated by the leading terms of *all* the polynomials in I, then $\{g_1, \ldots, g_k\}$ is a Gröbner basis. (In fact, this is often taken as the definition of a Gröbner basis.)

Finding Gröbner bases

15.20   Using the notation from the proof of Theorem 15.7 prove that, if the remainder is 0 when any S-polynomial $S(g_{i_s}, g_{i_t})$ is divided by the polynomials $g_1, \ldots, g_k$, then the remainder is also 0 when the S-polynomial

$$S(x_1^{b_{i_s 1}} x_2^{b_{i_s 2}} \cdots x_n^{b_{i_s n}} g_{i_s}, x_1^{b_{i_t 1}} x_2^{b_{i_t 2}} \cdots x_n^{b_{i_t n}} g_{i_t})$$

is divided by the $g_1, \ldots, g_k$.

15.21   Verify that the S-polynomial $xz + yz$ from Example 15 has a remainder 0 when divided by the polynomials $x + z$ and $y - z$.

15.22 Use Theorem 15.7 to verify that the set $\{x^2, xy + y^2, y^3\}$ is a Gröbner basis for the ideal $I = (x^2, xy + y^2, y^3)$ in $\mathbf{R}[x, y]$ with the lexicographic order where $x > y$.

15.23 Does the set $\{y - x^2, z - x^3\}$ form a Gröbner basis for the ideal $I = (y - x^2, z - x^3)$ in $\mathbf{R}[x, y, z]$? Consider this question first with the lexicographic order where $x > y > z$, and then consider it again with the lexicographic order where $z > y > x$.

15.24 In Example 1 it was claimed that

$$\{x^2y - y + x, -y^2 + xy + x^2, x^3 + y - 2x\}$$

is a reduced Gröbner basis for the ideal $I = (x^2y - y + x, x^2y - y + x)$ in the ring $\mathbf{Q}[x, y]$. Verify that this set is indeed a Gröbner basis using Theorem 15.7, and concluded that it is a reduced Gröbner basis.

15.25 Example 1 showed that the polynomial

$$x^5 - x^4y - 2x^3y^2 + 2x^2y^3 - 2xy^4 - y^5 + 2x^3 - 3x^2y + 4xy^2 + y^3 - x - y$$

is in the ideal $I = (xy^2 - x + y, x^2y - y)$ in the ring $\mathbf{Q}[x, y]$, but omitted the details. Fill in those details now by actually constructing the reduced Gröbner basis for I, and then using the division algorithm to express this polynomial in terms of that basis. Finally, express the polynomial in terms of the two original generators of the ideal I.

Solutions to Selected Problems

Chapter 1

1.1   Let 0 be the zero element in R. Suppose there is another element $0'$ that behaves like 0 — that is, an element such that, for all $a \in R$, $a + 0' = a$. Then, $0 = 0 + 0' = 0' + 0 = 0'$.

   Let 1 be the multiplicative identity. Suppose there is another element $1'$ that behaves like 1 — that is, an element such that, for all $a \in R$, $a \cdot 1' = a$. Then, $1 = 1 \cdot 1' = 1' \cdot 1 = 1'$.

   Let $a \in R$. Suppose there is another element b that behaves like $-a$ — that is, an element such that $a + b = 0$. Then $-a = 0 + (-a) = (a + b) + (-a) = (b + a) + (-a) = b + (a + (-a)) = b + 0 = b$.

1.2   (1) $0a + 0a = (0 + 0)a = 0a$. We can subtract $0a$ from each side — that is, add $-0a$ to each side — leaving $0a = 0$.

   (2) Since the element ab has only one additive inverse (by Problem 1.1), we will show that $(-a)b$ is that inverse by showing that $ab + ((-a)b) = 0$. The details of this computation are $ab + ((-a)b) = (a + (-a))b = 0b = 0$.

   (3) $a(b - c) = a(b + (-c)) = ab + a(-c) = ab + (-(ac)) = ab - ac$.

1.3   We use Problem 1.2: $(-a)(-b) = -((-a)b) = -(-(ab)) = ab$.

1.4   (i) $(ab)^n = a^n b^n$: This is trivially true if $n = 0$. Assume that it is true for $n = k - 1$. Then

$$(ab)^k = (ab)^{k-1}(ab) = a^{k-1}b^{k-1}ab = a^{k-1}ab^{k-1}b = a^k b^k.$$

Thus, by induction, the formula holds for all n.

   (ii) $a^m a^n = a^{m+n}$: We fix m. The formula is trivially true if $n = 0$. Assume it is true for $n = k - 1$. Then

$$a^m a^k = a^m a^{k-1} a = a^{m+k-1} a = a^{m+k}.$$

Thus, by induction, the formula holds for all n and hence for all m and n.

   (iii) $(a^m)^n = a^{mn}$: We fix m. The formula is trivially true if $n = 0$. Assume it is true for $n = k - 1$. Then

$$(a^m)^k = (a^m)^{k-1}(a^m) = a^{m(k-1)}a^m = a^{m(k-1)+m} = a^{mk}.$$

Thus, by induction, the formula holds for all n and hence for all m and n.

1.5   Let a be an element of the ring R. Then $a = a \cdot 1 = a \cdot 0 = 0$. Therefore, there is only one element in the ring — namely, 0.

1.7   Axiom 1: addition is associative, since

$$(a+bi)+((c+di)+(e+fi)) = (a+bi)+((c+e)+(d+f)i)$$
$$= (a+(c+e))+(b+(d+f))i = ((a+c)+e)+((b+d)+f)i$$
$$= ((a+c)+(b+d)i)+(e+fi) = ((a+bi)+(c+di))+(e+fi).$$

Axiom 2: addition is commutative, since

$$(a+bi)+(c+di) = (a+c)+(b+d)i = (c+a)+(d+b)i$$
$$= (c+di)+(a+bi).$$

Axiom 3: the zero element is $0+0i$, since

$$(a+bi)+(0+0i) = a+bi.$$

Axiom 4: the (additive) inverse of $a+bi$ is $-a+(-b)i$, since

$$(a+bi)+(-a+(-b)i) = 0+0i.$$

Axiom 5: multiplication is associative, since

$$(a+bi)((c+di)(e+fi)) = (a+bi)((ce-df)+(cf+de)i)$$
$$= (a(ce-df)-b(cf+de))+(a(cf+de)+b(ce-df))i$$
$$= ((ac-bd)e-(ad+bc)f)+((ac-bd)f+(ad+bc)e)i$$
$$= ((ac-bd)+(ad+bc)i)(e+fi)$$
$$= ((a+bi)(c+di))(e+fi).$$

Axiom 6: multiplication is distributive over addition, since

$$((a+bi)+(c+di))(e+fi) = ((a+c)+(b+di))(e+fi)$$
$$= ((a+c)e-(b+d)f)+((a+c)f+(b+d)e)i$$
$$= (ae+ce-bf-df)+(af+cf+be+de)i$$
$$= ((ae-bf)+(ce-df))+((af+be)+(cf+de))i$$
$$= ((ae-bf)+(af+be)i)+((ce-df)+(cf+de)i)$$
$$= (a+bi)(e+fi)+(c+di)(e+fi).$$

Actually, we have only shown that multiplication distributes over addition from the right, but the proof of the "from the left" case is similar. Alternatively, the remaining case follows easily from commutivity, which is Axiom 3.

Axiom 7: multiplication is commutative, since

$$(a + bi)(c + di) = (ac - bd) + (ad + bc)i = (ca - db) + (cb + da)i$$
$$= (c + di)(a + bi).$$

Axiom 8: the multiplicative identity is $1 + 0i$, since

$$(1 + 0i)(a + bi) = (1a - 0b) + (1b + 0a)i = a + bi.$$

1.8   No. The set of even integers is closed under addition and multiplication, and contains additive inverses, but does not contain 1, the identity of $\mathbf{Z}$.

1.9   Yes. If you add or multiply two rational numbers you get a rational number; the negative of a rational number is a rational number; and 1, the identity of $\mathbf{R}$, is a rational number.

Chapter 2

2.7   Let $a, b \in I \cap J$; then $a, b \in I$ and $a, b \in J$, so $a - b \in I$ and $a - b \in J$ since I and J are ideals of R. Thus, $a - b \in I \cap J$, as desired.
      Now, let $r \in R$ and $a \in I \cap J$; then $a \in I$ and $a \in J$, and since I and J are both ideals of R, it follows that $ra \in I$ and $ra \in J$. Thus, $ra \in I \cap J$, as desired and $I \cap J$ is an ideal of R.

2.13   Suppose that $a \in I$, where I is an ideal. Take a typical element ra of the ideal (a), where $r \in R$. Since $a \in I$ and I is an ideal, $ra \in I$. Therefore, $(a) \subseteq I$, as desired.

2.14   As usual, in order to prove equality, we demonstrate containment in both directions. Problem 2.13 says that the ideal (a) is contained in any ideal containing a, and so (a) must be contained in the intersection of all ideals containing a.
      In the other direction, (a) is itself an ideal containing a and, as such, is one of the ideals in the intersection. Therefore, the intersection is contained in (a). Thus, we have demonstrated containment in both directions, and (a) equals the intersection.

2.15   Let $r_1 a + s_1 b$ and $r_2 a + s_2 b$ be two elements of (a, b), where $r_i, s_i \in R$. Then

$$(r_1 a + s_1 b) - (r_2 a + s_2 b) = (r_1 - r_2)a + (s_1 - s_2)b \in (a, b).$$

Next, let $r \in R$. Then

$$r(r_1 a + s_1 b) = (rr_1)a + (rs_1)b \in (a, b).$$

Thus (a, b) is an ideal.

2.20   First, assume that R is a field. Let I be an ideal of R. We must show that $I = 0$ or $I = R$. If $I \neq 0$, then I contains some element $a \neq 0$. Since R is a field, the element a has an inverse a^{-1}. But, $1 = a^{-1}a \in I$, since I is an ideal and $a \in I$. Hence, $1 \in I$ and $I = R$ (see Example 2 on page 12). Therefore, the only ideals of R are 0 and R.

Conversely, assume that the only ideals of R are 0 and R. Let $a \neq 0 \in R$. Then, since (a) is a nonzero ideal, $(a) = R$. In particular, this means that $1 \in (a)$. Thus, there is an element $r \in R$, such that $1 = ra$. Therefore, the non-zero element a has an inverse (namely r), and R is a field.

2.21   Since the sum and the product of two continuous functions are continuous, $C(\mathbf{R})$ is closed under addition and multiplication.

Addition is associative, since if $f, g, h \in C(\mathbf{R})$, then

$$((f + g) + h)(x) = (f + g)(x) + h(x) = (f(x) + g(x)) + h(x)$$
$$= f(x) + (g(x) + h(x)) = f(x) + (g + h)(x) = (f + (g + h))(x).$$

Addition is commutative, since

$$(f + g)(x) = f(x) + g(x) = g(x) + f(x) = (g + f)(x).$$

The zero element is 0 — that is, the function such that $0(x) = 0$ for all x — since

$$(0 + f)(x) = 0(x) + f(x) = 0 + f(x) = f(x).$$

Each element $f \in C(\mathbf{R})$ has an additive inverse, $-f$, the function defined by $(-f)(x) = -(f(x))$, since

$$(f + (-f))(x) = f(x) + (-f)(x) = f(x) - f(x) = 0,$$

and therefore $f + (-f) = 0$.

Multiplication is associative, since if $f, g, h \in C(\mathbf{R})$, then

$$(f(gh))(x) = f(x)(gh)(x) = f(x)(g(x)h(x)) = (f(x)g(x))h(x)$$
$$= (fg)(x)h(x) = ((fg)h)(x).$$

Multiplication is distributive over addition, since if $f, g, h \in C(\mathbf{R})$, then

$$((f + g)h)(x) = ((f + g)(x))(h(x)) = (f(x) + g(x))(h(x))$$
$$= f(x)h(x) + g(x)h(x) = (fh)(x) + (gh)(x) = (fh + gh)(x).$$

Multiplication is commutative, since

$$(fg)(x) = f(x)g(x) = g(x)f(x) = (gf)(x).$$

The identity element is 1 — that is, the function such that $1(x) = 1$ for all x — since

$$(1f)(x) = 1(x)f(x) = 1f(x) = f(x).$$

Therefore, $C(\mathbf{R})$ is a ring.

2.22   If $f, g \in C(\mathbf{R})$), then $f(17) = 0$ and $g(17) = 0$. So $(f - g)(17) = f(17) - g(17) = 0 - 0 = 0$, and $f - g \in C(\mathbf{R})$). If $f \in C(\mathbf{R})$) and $g \in I$, then g(17)=0. So $(fg)(17) = f(17)g(17) = f(17) \cdot 0 = 0$, and $fg \in C(\mathbf{R})$). Therefore, I is an ideal.

2.23   We use basic property (2) of cosets. We begin with $a + I = a' + I$ and $b + I = b' + I$. Thus, $a - a' \in I$ and $b - b' \in I$. Since I is an ideal, it follows that $(a - a') + (b - b') \in I$. In other words, $(a + b) - (a' + b') \in I$. Therefore, $(a + b) + I = (a' + b') + I$, and addition of cosets is well defined.
    We now show that $ab + I = a'b' + I$. In order to do this, we show that $ab - a'b' \in I$. But

$$ab - a'b' = a(b - b') + (a - a')b' \in I,$$

since $b - b' \in I$ and $a - a' \in I$. Therefore, multiplication of cosets is also well defined.

2.24   Addition is associative, since

$$(a + I) + \big((b + I) + (c + I)\big) = (a + I) + (b + b + I)$$
$$= a + (b + c) + I = (a + b) + c + I$$
$$= (a + b + I) + (c + I) = \big((a + I) + (b + I)\big) + (c + I).$$

Addition is commutative, since

$$(a + I) + (b + I) = a + b + I = b + a + I = (b + I) + (a + I).$$

The zero element is $0 + I = I$, since $(0 + I) + (a + I) = a + I$.
The additive inverse of an element $a + I$ is $-a + I$, since

$$(a + I) + (-a + I) = a + (-a) + I = 0 + I = I.$$

Multiplication is associative, since

$$(a + I)\big((b + I)(c + I)\big) = (a + I)(bc + I) = a(bc) + I$$
$$= (ab)c + I = (ab + I)(c + I) = \big((a + I)(b + I)\big)(c + I).$$

Multiplication is distributive over addition, since

$$\big((a + I) + (b + I)\big)(c + I) = (a + b + I)(c + I)$$
$$= (a + b)c + I = ac + bc + I$$
$$= (ac + I) + (bc + I) = (a + I)(c + I) + (b + I)(c + I).$$

Multiplication is commutative, since

$$(a + I)(b + I) = ab + I = ba + I = (b + I)(a + I).$$

The identity of R/I is $1 + I$, since $(a + I)(1 + I) = a + I$. Therefore, all of the axioms are satisfied and R/I is a ring.

2.25   First, assume that n is a prime number. We must show that any nonzero
        element of $\mathbf{Z}/(n)$ has an inverse. Let $m + (n)$ be a nonzero element of $\mathbf{Z}/(n)$ —
        that is, $m + (n) \neq (n)$. This means that $m \notin (n)$ — that is, m is not a multiple of n.
        Since n is prime, we conclude that the greatest common divisor of m and n is 1.
        Using, for example, the Euclidean algorithm, we can write $1 = rm + sn$ for
        some $r, s \in \mathbf{Z}$. We claim that $r + (n)$ is the inverse of $m + (n)$. This follows
        immediately since

$$\left(r + (n) \right) \left(m + (n) \right) = rm + (n) = 1 - sn + (n) = 1 + (n).$$

        Thus, $\mathbf{Z}/(n)$ is a field.
            Conversely, assume that $\mathbf{Z}/(n)$ is a field. We show that n is a prime number.
        We give a proof by contradiction. Suppose that n is not prime, then we can
        write $n = ab$, where $a, b \in \mathbf{Z}$ and $a, b > 1$. We claim that the nonzero element
        $a + (n)$ does not have an inverse. For if, say, $a' + (n)$ is the inverse of $a + (n)$, then

$$1 + (n) = \left(a' + (n) \right) \left(a + (n) \right) = a'a + (n).$$

        It follows that

$$b + (n) = \left(1 + (n) \right) \left(b + (n) \right)$$
$$= \left(a'a + (n) \right) \left(b + (n) \right) = a'ab + (n) = a'n + (n) = (n).$$

        Therefore, $b \in (n)$; however, this is impossible, since b was chosen to be a proper
        divisor of n, so that $1 < b < n$. We conclude that the nonzero element $a + (n)$
        does not have an inverse, and, as a result, that $\mathbf{Z}/(n)$ is not a field. This
        contradicts the hypothesis, and so, n is a prime number after all.

Chapter 3

3.1   Since $(ab)c \in P$ and P is prime, then $ab \in P$ or $c \in P$; but, if $ab \in P$, then $a \in P$
        or $b \in P$; thus, $a \in P$, $b \in P$, or $c \in P$.
            For the general case we use induction. The case $n = 2$ is precisely the
        definition of prime ideal (and, by the way, we just did the case $n = 3$). Assume
        that the statement is true for $n = k - 1$. Then

$$a_1 a_2 \cdots a_k = (a_1 a_2 \cdots a_{k-1})a_k \in P,$$

        so, since P is prime, $a_1 a_2 \cdots a_{k-1} \in P$ or $a_k \in P$. If $a_k \in P$, then we are done.
        Otherwise, $a_1 a_2 \cdots a_{k-1} \in P$, and, by the induction hypothesis, one of the
        elements $a_1, a_2, \ldots, a_{k-1}$ is in P, as desired.

3.2   First, assume that (0) is a prime ideal. We must show that R is a domain.
        Suppose that $a \in R$ is a zero-divisor, then $ab = 0$, for some $b \neq 0 \in R$. Thus,
        $ab \in (0)$, but $b \notin (0)$. Therefore, since (0) is prime, $a \in (0)$. This means that
        $a = 0$. We conclude that the only zero-divisor in R is 0 — that is, R is an
        integral domain.

Conversely, assume that R is an integral domain. We must show that (0) is a prime ideal. Let $ab \in (0)$, where $a, b \in R$. Then, $ab = 0$, so $a = 0$ or $b = 0$ (since R is an integral domain). Therefore, $a \in (0)$ or $b \in (0)$, and (0) is a prime ideal, as desired.

3.3   By definition, the ideal P is prime *if and only if* whenever $ab \in P$, then $a \in P$ or $b \in P$, but this is true *if and only if* whenever $a \notin P$ and $b \notin P$, then $ab \notin P$, which is true *if and only if* whenever $a \in R \setminus P$ and $b \in R \setminus P$, then $ab \in R \setminus P$, which is the statement that $R \setminus P$ is multiplicatively closed.

3.6   Since neither I nor J contains the other, there is an element $a \in I$ with $a \notin J$, and an element $b \in J$ with $b \notin I$. Then $ab \in I$ (since $a \in I$), and $ab \in J$ (since $b \in J$). Therefore, $ab \in I \cap J$. However, $a \notin I \cap J$ and $b \notin I \cap J$, so $I \cap J$ is not a prime ideal.

3.8   Let $i_1 + r_1 a$ and $i_2 + r_2 a$ be two elements of (I, a), where $i_k \in I$ and $r_k \in R$. Then

$$(i_1 + r_1 a) - (i_2 + r_2 a) = (i_1 - i_2) + (r_1 - r_2)a \in (I, a).$$

Next, let $r \in R$. Then

$$r(i_1 + r_1 a) = r i_1 + (r r_1)a \in (I, a).$$

Thus, (I, a) is an ideal.

3.9   Let M be a maximal ideal of a ring R. Let $a, b \in R$ such that $ab \in M$. We must show that $a \in M$ or $b \in M$. Assume that $a \notin M$. Then the ideal (M, a) is strictly larger than M, but M is maximal, and so, $(M, a) = R$. In particular, $1 \in (M, a)$, so we can write $1 = m + ra$, for some $m \in M$ and some $r \in R$. Therefore, multiplying by b, we have $b = bm + rab \in M$, since $ab \in M$. Thus, $b \in M$, as desired. Therefore, M is a prime ideal.

3.10   We claim that (0) is a maximal ideal that is not prime. The ideal (0) is not prime, since $2 \cdot 3 = 0 \in (0)$, but $2 \notin (0)$ and $3 \notin (0)$. However, (0) is maximal, since the only other ideal is the entire ring R.

3.11   Let $a_1, a_2, \ldots, a_n$ be *all* of the elements of D. We must show that any nonzero element a of D has an inverse. We claim that all the products $aa_1, aa_2, \ldots, aa_n$ are distinct. If not, suppose that $aa_i = aa_j$ for some $i \neq j$. Therefore, $a(a_i - a_j) = 0$, but D is an integral domain and $a \neq 0$, so we have $a_i - a_j = 0$. Thus, $a_i = a_j$, which is a contradiction since $i \neq j$. We conclude that the elements $aa_1, aa_2, \ldots, aa_n$ are distinct. But D has only n elements, one of which is 1, the identity. Therefore, one of the n products, say aa_i, must be the element 1, and, hence, a_i is the inverse of a. We conclude that D is a field, since any nonzero element has an inverse.

3.12   Let P be a prime ideal of R. Then, by Theorem 3.1, R/P is an integral domain; also R/P is certainly finite (since R is finite). Hence, by Problem 3.11, R/P must be a field, and, by Theorem 3.2, we can then conclude that P is a maximal ideal.

3.13   First, we assume that P is prime. Let I and J be two ideals of R such that $IJ \subseteq P$. We must show that $I \subseteq P$ or $J \subseteq P$. Suppose that $I \nsubseteq P$ — that is, there is an element $a \in I$ such that $a \notin P$. Then, for *every* $b \in J$, $ab \in IJ \subseteq P$. Thus, for *all* $b \in J$, $ab \in P$, but $a \notin P$, so $b \in P$. Therefore, $J \subseteq P$.

     Conversely, assume that $IJ \subseteq P$ implies $I \subseteq P$ or $J \subseteq P$, for any ideals I and J of R. We must show that P is prime. Let $a, b \in R$ such that $ab \in P$. Then the product of the principal ideals (a) and (b) is contained in P — that is, $(a)(b) \subseteq P$. Therefore, by hypothesis, $(a) \subseteq P$ or $(b) \subseteq P$. We conclude that $a \in P$ or $b \in P$, and that P is prime.

3.14   We do the easier half first. Assume that $D[x]$ is an integral domain. But D is a subring of $D[x]$, so D is also an integral domain (since any zero-divisor in D must also be a zero-divisor in $D[x]$).

Conversely, assume that D is a domain. We give a proof by contradiction. Suppose that $f \neq 0 \in D[x]$ is a zero-divisor. Let g be a nonzero polynomial in $D[x]$ such that $fg = 0$. We write $f = a_0 + a_1 x + \cdots + a_n x^n$ and $g = b_0 + b_1 x + \cdots + b_m x^m$, where $a_i, b_i \in D$, and $a_n, b_m \neq 0$. Since $fg = 0$, we get

$$0 = fg = a_0 b_0 + (a_0 b_1 + a_1 b_0)x + \cdots + a_n b_m x^{n+m},$$

Therefore, each of the coefficients of this product must be 0. In particular, $a_n b_m = 0$. But, neither a_n nor b_m equals 0, which contradicts the fact that D is an integral domain. We conclude that $D[x]$ has no zero-divisors (other than 0), and so $D[x]$ is an integral domain.

3.15   Since any polynomial in x and y with coefficients in D can be thought of as a polynomial in y with coefficients in $D[x]$, we can think of the ring $D[x, y]$ as the ring $(D[x])[y]$. Hence, we can apply Problem 3.14 twice: D is an integral domain if and only if $D[x]$ is an integral domain if and only if $(D[x])[y]$ is an integral domain.

The general case follows in exactly the same way since the ring $D[x_1, x_2, \ldots, x_n]$ is the same as the ring $\big(D[x_1, x_2, \ldots, x_{n-1}]\big)[x_n]$.

Chapter 4

4.1   Let I be an ideal of a ring R. Let S be the set of *all* proper ideals of R that contain S. Then S is a partially ordered set under inclusion. Note that S is non-empty since I itself is a proper ideal of R that contains I. We use Zorn's lemma to find a maximal element of S.

Let C be a chain in S — that is, C is a chain of proper ideals each containing I. Let U be the union of *all* the ideals in the chain C. First, we show that U is an ideal.

Let $x, y \in U$. Then $x \in I_1$ and $y \in I_2$ for some ideals I_1 and I_2 in the chain C. But, by definition, the chain C is totally ordered, so I_1 and I_2 are comparable. This means that, say, $I_1 \subseteq I_2$. Therefore, $x \in I_2$, and so, $x - y \in I_2$, since I_2 is an ideal. Thus, $x - y \in U$, since U is the union of all the ideals in the chain C and, as such, contains I_2.

Now let $x \in U$ and let $r \in R$. Then $x \in I$ for some ideal I in the chain C. Thus, $rx \in I$, since I is an ideal, and so $rx \in U$. Therefore, U is an ideal of R.

Next, U is a proper ideal of R since $1 \notin U$. This is because 1 is not in any of the ideals in the chain C, since all these ideals are themselves proper ideals.

Therefore, U is an upper bound in S for the chain C. So, by Zorn's lemma, S has a maximal element, which, by the definition of the set S, must also be a maximal ideal containing I.

4.2   This, of course, is another form of Russell's paradox. If "heterological" is autological, then it applies to itself, and so "heterological" is heterological (rather than autological as we assumed). On the other hand, if "heterological" is heterological, then it does not apply to itself, and so "heterological" is not heterological — that is, it is autological (rather than heterological as we assumed).

4.3   If $S \in S$, then, by definition of S, S is not a member of itself — that is, $S \notin S$, a contradiction. On the other hand, if $S \notin S$, then, by definition of S, S is a member of itself — that is, $S \in S$, a contradiction.

4.4   Let $a, b \in R$ be such that $ab \in \bigcap P_\lambda$. Then $ab \in P_\lambda$ for each λ in the index set Λ. We must show that $a \in P_\lambda$ for each λ, or that $b \in P_\lambda$ for each λ. Suppose, then, that there is a $\gamma \in \Lambda$ such that $a \notin P_\gamma$. Then $b \in P_\gamma$, since P_γ is prime. We must show that $b \in P_\lambda$ for each λ. Since $\{P_\lambda\}$ is a chain, either $P_\lambda \subseteq P_\gamma$ or $P_\gamma \subseteq P_\lambda$ for each λ. If $P_\gamma \subseteq P_\lambda$, then $b \in P_\gamma$, and so $b \in P_\lambda$. On the other hand, if $P_\lambda \subseteq P_\gamma$, then $a \notin P_\lambda$ (since, otherwise, $a \in P_\lambda \subseteq P_\gamma$, which is a contradiction); therefore, since P_λ is prime, $b \in P_\lambda$. We conclude that $\bigcap P_\lambda$ is prime.

4.5   Let S be the set of all prime ideals that contain I *and* are contained in Q. Then Q is in S, so the set S is non-empty. Order S by reverse inclusion — that is, if $A, B \in S$, then $A \leq B$ if and only if $B \subseteq A$. In order to apply Zorn's lemma, we must show that any chain in S has an upper bound in S. Let C be a chain in S. In particular, the elements of the chain C are prime ideals containing I and contained in Q. We claim that the intersection J of this chain is an upper bound. By Problem 4.4, J is a prime ideal. Since each of the ideals in the chain contains I and is contained in Q, J contains I and is contained in Q. Thus, J is in S. Moreover, J is an upper bound for the chain since it is contained in each ideal of the chain. Therefore, by Zorn's lemma, S has a maximal element. But, since we are using reverse inclusion, that maximal element of S is the minimal prime ideal we are seeking.

4.6   Let R be a ring, let $I = (0)$, and let Q be a maximal ideal of R — such a Q exists by Theorem 4.2. Then, by Problem 4.5, there is a minimal prime ideal P over (0) — that is, a minimal prime ideal.

Chapter 5

5.6   Since a is nilpotent, $a^n = 0$ for some positive integer n. Then
        $1 - a + a^2 - a^3 + \cdots + (-1)^{n-1}a^{n-1}$ is the inverse of $1 + a$; hence, $1 + a$ is a unit.

5.8   Let $a_1 + b_1, a_2 + b_2 \in I + J$, where $a_i \in I$ and $b_i \in J$. Then

$$(a_1 + b_1) - (a_2 + b_2) = (a_1 - a_2) + (b_1 - b_2) \in I + J,$$

since $a_1 - a_2 \in I$ and $b_1 - b_2 \in J$.
        Let $r \in R$ and $a + b \in I + J$, where $a \in I$ and $b \in J$. Then

$$r(a + b) = ra + rb \in I + J,$$

since $ra \in I$ and $rb \in J$. Thus, $I + J$ is an ideal of R.
        Next, we show that if $x \in R$, then $I + (x) = (I, x)$. But a typical element of $I + (x)$ is of the form $i + rx$, where $i \in I$ and $rx \in (x)$, which is exactly the form for elements of (I, x).

5.9 It suffices to show that the product of an element from $I + (x)$ and an element from $I + (y)$ is in $I + (xy)$. Thus, let $i_1 + r_1 x \in I + (x)$ and let $i_2 + r_2 y \in I + (y)$, where $i_k \in I$ and $r_k \in R$. Then

$$(i_1 + r_1 x) \cdot (i_2 + r_2 y) = (i_1 i_2 + r_1 x i_2 + r_2 y i_1) + (r_1 r_2) \cdot (xy),$$

and so the product is an element of $I + (xy)$, as desired.

In order to show that we need not have equality for these two sets, let K be a field and consider the ring $K[x]$ of polynomials over K. Let $I = (x)$ — that is, the ideal of polynomials with zero constant term. Let $a = b = 0$. Then $I + (a) = I + (b) = I$, and the product of these two ideals is (x^2). However, $I + (ab) = I$, which is (x). Clearly, $(x^2) \neq (x)$, since $x \notin (x^2)$.

Chapter 6

6.2 Addition is associative, since

$$[a/b] + ([c/d] + [e/f]) = [a/b] + [(cf + de)/bd]$$
$$= [(adf + bcf + bde)/bdf] = [(ad + bc)/bd] + [e/f]$$
$$= ([a/b] + [c/d]) + [e/f].$$

Addition is commutative, since

$$[a/b] + [c/d] = [(ad + bc)/bd] = [(cb + da)/db] = [c/d] + [a/b].$$

The zero element is $[0/1]$, since

$$[a/b] + [0/1] = [(a \cdot 1 + b \cdot 0)/b \cdot 1] = [a/b].$$

The additive inverse of $[a/b]$ is $[-a/b]$, since

$$[a/b] + [-a/b] = [(ab - ba)/b^2] = 0.$$

Multiplication is associative, since

$$[a/b] \cdot ([c/d] \cdot [e/f]) = [a/b] \cdot [ce/df] = [ace/bdf]$$
$$= [ac/bd] \cdot [e/f] = ([a/b] \cdot [c/d]) \cdot [e/f].$$

Multiplication is distributive over addition, since

$$([a/b] + [c/d]) \cdot [e/f] = [(ad + bc)/bd] \cdot [e/f]$$
$$= [(ade + bce)/bdf] = [(aedf + bfce)/bfdf]$$
$$= [ae/bf] + [ce/df] = [a/b] \cdot [e/f] + [c/d] \cdot [e/f].$$

Multiplication is commutative, since

$$[a/b] \cdot [c/d] = [ac/bd] = [ca/db] = [c/d] \cdot [a/b].$$

The multiplicative identity is $[1/1]$, since

$$[1/1] \cdot [a/b] = [1 \cdot a/1 \cdot b] = [a/b].$$

Therefore, F is a ring. This proof does not change at all when verifying that the total quotient ring is a ring.

6.6   Suppose that $[a/b] = [a'/b']$ and $[c/d] = [c'/d']$. We must show that $[a/b] + [c/d] = [a'/b'] + [c'/d']$, and that $[a/b] \cdot [c/d] = [a'/b'] \cdot [c'/d']$. We know that $s(ab' - ba') = 0$ for some $s \in T$, and that $t(cd' - dc') = 0$ for some $t \in T$. So,

$$st\big((ad + bc)b'd' - bd(a'd' + b'c')\big)$$
$$= st(adb'd' + bcb'd') - st(bda'd' + bdb'c')$$
$$= s(ab' - ba')tdd' + t(cd' - dc')sbb' = 0.$$

Thus, since $st \in T$, $[(ad + bc)/bd] = [(a'd' + b'c')/b'd']$, as desired, and addition is well defined.

Similarly,

$$st(acb'd' - bda'c') = st(ab'cd' - ba'cd' + cd'ba' - dc'ba')$$
$$= s(ab' - ba')tcd' + t(cd' - dc')sba' = 0.$$

Thus, since $st \in T$, $[ac/bd] = [a'c'/b'd']$, and multiplication is well defined.

6.7   Addition is associative, since

$$[a/b] + \big([c/d] + [e/f]\big) = [a/b] + [(cf + de)/df]$$
$$= [(adf + bcf + bde)/bdf] = [(ad + bc)/bd] + [e/f]$$
$$= \big([a/b] + [c/d]\big) + [e/f].$$

Addition is commutative, since

$$[a/b] + [c/d] = [(ad + bc)/bd]$$
$$= [(cb + da)/db] = [c/d] + [a/b].$$

The zero element is $[0/1]$, since

$$[a/b] + [0/1] = [(a \cdot 1 + b \cdot 0)/b \cdot 1] = [a/b].$$

The additive inverse of $[a/b]$ is $[-a/b]$, since

$$[a/b] + [-a/b] = [(ab - ba)/b^2] = 0.$$

Multiplication is associative, since

$$[a/b] \cdot ([c/d] \cdot [e/f]) = [a/b] \cdot [ce/df] = [ace/bdf]$$
$$= [ac/bd] \cdot [e/f] = ([a/b] \cdot [c/d]) \cdot [e/f].$$

Multiplication is distributive over addition, since

$$([a/b] + [c/d]) \cdot [e/f] = [(ad + bc)/bd] \cdot [e/f]$$
$$= [(ade + bce)/bdf] = [(aedf + bfce)/bfdf]$$
$$= [ae/bf] + [ce/df] = [a/b] \cdot [e/f] + [c/d] \cdot [e/f].$$

Multiplication is commutative, since

$$[a/b] \cdot [c/d] = [ac/bd] = [ca/db] = [c/d] \cdot [a/b].$$

The multiplicative identity is $[1/1]$, since

$$[1/1] \cdot [a/b] = [1 \cdot a/1 \cdot b] = [a/b].$$

Therefore, R_T is a ring.

6.8   If D is an integral domain and T is the set of nonzero elements of D, then D_T is the quotient field of D. This is because the equivalence relation for the localization with respect to T is given by

$$a/b \cong c/d \text{ if and only if } ad = bc,$$

since D is an integral domain. So, $t(ad - bc) = 0$ implies that $ad - bc = 0$.
      Similarly, for an arbitrary ring R, if we let T be the set of regular elements of R, then R_T is the total quotient ring of R.

6.9   First, assume that $0 \in T$, and let $a/b \in R_T$. We must show that $a/b = 0/1$. But, $0(a \cdot 1 - b \cdot 0) = 0$, and $0 \in T$, so $a/b = 0/1$. Therefore, $R_T = 0$.
      Conversely, if $R_T = 0$, then $1/1 = 0/1$. Thus, there is an element $t \in T$ such that $t(1 \cdot 1 - 1 \cdot 0) = 0$ — namely, $t = 0$. Therefore, $0 \in T$.

6.10 $R_T = R$, since, if u is a unit of R, then any element $a/u \in R_T$ corresponds to the element $au^{-1} \in R$. In other words, we can already divide by u in R, and so nothing is added when we "include" $1/u$ in R_T.

6.11 $\mathbf{Z}_T$ just consists of all rational numbers whose denominators are a power of n.
      In an arbitrary ring R, as long as $a \in R$ is not a nilpotent element, then we can localize at the powers of a, since otherwise, $a^n = 0$, for some n, and then the localization would collapse to 0 — that is, to the zero ring (see Problem 6.9).

6.13   Clearly, T is a multiplicative system. $K[x]_T$ consists of all rational functions f/g such that $g(a) \neq 0$. In particular, this means that we can "evaluate" f/g at a.

6.14   Let R be a ring between $\mathbf{Z}$ and $\mathbf{Q}$, and let T be the set of all integers t such that R contains the rational number $1/t$. Clearly, T is a multiplicative system of $\mathbf{Z}$.

If $a/t \in \mathbf{Z}_T$, where $a \in \mathbf{Z}$ and $t \in T$, then $1/t \in R$. Therefore, $a/t = a(1/t) \in R$. Thus, $\mathbf{Z}_T \subseteq R$.

If $a/b \in R$, where $a, b \in \mathbf{Z}$, then we can assume that a and b are relatively prime. This means that there are integers r and s such that $1 = ra + sb$. Therefore, $1/b = r(a/b) + s \in R$. It follows that $b \in T$, and so, $a/b \in \mathbf{Z}_T$. We conclude that $R = \mathbf{Z}_T$.

6.17 $\mathbf{Z}_{(p)}$ consists of all rational numbers a/b, where b is relatively prime to p — that is, b is not divisible by p. Since (p) is a prime ideal, Example 14 applies, and $\mathbf{Z}_{(p)}$ is a local ring. The unique maximal ideal is $(p)_{(p)}$ — that is, the set of all rational numbers a/b where a is a multiple of p and b is relatively prime to p. (Note that all other elements are units in this ring.)

6.19   Let $r/s \in \mathcal{N}_T$, where $r \in \mathcal{N}$ and $s \in T$. Since $r \in \mathcal{N}$, $r^n = 0$ for some positive integer n. So, $(r/s)^n = 0$, and r/s is an element of the nilradical of R_T, as desired.

On the other hand, if $r/s \in R_T$ is an element of the nilradical of R_T, then $(r/s)^n = 0$ for some positive integer n — that is, $r^n/s^n = 0$. This means that $tr^n = 0$ for some $t \in T$. Therefore, $(tr)^n = 0$, and so, $tr \in \mathcal{N}$. Thus, $r/s = tr/ts \in \mathcal{N}_T$, as desired. We conclude that $\mathcal{N}_T$ is the nilradical of R_T.

6.20   Let R be a ring with no nonzero nilpotent elements. Then the nilradical of R is (0). So, by Problem 6.19, if P is a prime ideal, then the nilradical of R_P is $(0)_P$, which is (0). Therefore, R_P also has no nonzero nilpotent elements.

Conversely, suppose that $a \in R$ is a nilpotent element of R. We must show that $a = 0$. Let $\mathrm{Ann}(a) = \{x \in R \mid xa = 0\}$. It is routine to check that the set $\mathrm{Ann}(a)$ is an ideal of R. (You may wonder about the unusual choice for a name for this ideal. It is an abreviation of the word "annihilator" and, in fact, this ideal is often called the **annihilator ideal** of a. The terminology seems unfortunate to say the least, but it is now standard and we are probably stuck with it. The point, of course, is that $\mathrm{Ann}(a)$ is the set of all elements that "annihilate" a.)

We will show that $\mathrm{Ann}(a) = R$. Suppose that $\mathrm{Ann}(a)$ is a proper ideal of R. Then, by Problem 4.1, $\mathrm{Ann}(a)$ is contained in a maximal, and hence, prime ideal P. Now, $a/1$ is a nilpotent element of R_P, but R_P has no nonzero nilpotent elements, so $a/1 = 0$ in R_P. Thus, there is an element $x \notin P$ such that $xa = 0$. This contradicts the fact that $\mathrm{Ann}(a) \subseteq P$. We conclude that $\mathrm{Ann}(a) = R$. Therefore, $1 \in \mathrm{Ann}(a)$, and so, $a = 1 \cdot a = 0$, as desired.

Chapter 7

7.2   (a) $Z(fg) = Z(f) \cup Z(g)$, since $(fg)(x) = 0$ if and only if either $f(x) = 0$ or $g(x) = 0$. $Z(f^2 + g^2) = Z(f) \cap Z(g)$, since $(f^2 + g^2)(x) = 0$ if and only if both $f(x) = 0$ and $g(x) = 0$.

(b) If $Z(f) \cap Z(g) = \emptyset$, then $f^2 + g^2$ is never zero by part (a). Thus, $f^2 + g^2$ is a unit contained in the ideal (f, g). Therefore, $(f, g) = C(X)$.

On the other hand, if $(f, g) = C(X)$, then we can write the identity function 1 as $1 = af + bg$, where $a, b \in C(X)$. Suppose that $x \in Z(f) \cap Z(g)$. Then $1(x) = a(x) f(x) + b(x) g(x) = 0$, since $f(x) = 0$ and $g(x) = 0$. This is a contradiction since $1(x) = 1$. Therefore, $Z(f) \cap Z(g) = \emptyset$.

7.3   Since it is always true that the product of two ideals is contained in their intersection, we need only prove that $P \cap Q \subseteq PQ$. If $f \in P \cap Q$, then $f \in P$ and $f \in Q$. Since f is a continuous real-valued function, so is $f^{1/3}$. Also $f^{1/3} f^{1/3} f^{1/3} = f$; thus, $f^{1/3} \in P$ and $f^{1/3} \in Q$, since P and Q are prime ideals. It follows that $f^{2/3} \in Q$, and so $f = f^{1/3} f^{2/3} \in PQ$, as desired.

7.4   We claim that the ideal (f) consists of all continuous functions g which are differentiable at $\frac{1}{2}$ and are 0 at $\frac{1}{2}$. In order to prove this, we first assume that g is such a function and define a function h on the interval $[0, 1]$ as follows:

$$h(x) = \frac{g(x)}{x - \frac{1}{2}}, \quad \text{for } x \neq \tfrac{1}{2}$$

and

$$h(x) = g'\left(\tfrac{1}{2}\right), \quad \text{for } x = \tfrac{1}{2}.$$

We can apply l'Hôpital's rule to get

$$\lim_{x \to \frac{1}{2}} h(x) = \lim_{x \to \frac{1}{2}} \frac{g(x)}{x - \frac{1}{2}} = \lim_{x \to \frac{1}{2}} \frac{g'(x)}{1} = g'\left(\tfrac{1}{2}\right),$$

and so h is continuous — that is, $h \in C([0, 1])$. Therefore, $g = hf \in (f)$.

Conversely, suppose that $g \in (f)$; then $g = hf$ for some $h \in C([0, 1])$. Obviously, $g(\frac{1}{2}) = 0$. We show that g is differentiable at $\frac{1}{2}$:

$$g'\left(\tfrac{1}{2}\right) = \lim_{\Delta x \to 0} \frac{g(\frac{1}{2} + \Delta x) - g(\frac{1}{2})}{\Delta x}$$

$$= \lim_{\Delta x \to 0} \frac{f(\frac{1}{2} + \Delta x)\, h(\frac{1}{2} + \Delta x) - f(\frac{1}{2})\, h(\frac{1}{2})}{\Delta x} = \lim_{\Delta x \to 0} \frac{\Delta x\, h(\frac{1}{2} + \Delta x)}{\Delta x}$$

$$= \lim_{\Delta x \to 0} h\left(\tfrac{1}{2} + \Delta x\right) = h\left(\tfrac{1}{2}\right),$$

since h is continuous at $\frac{1}{2}$. So, g is differentiable at $\frac{1}{2}$.

Next, we show that $(f)^2 \neq (f)$. In particular, we show that $f \notin (f)^2$. Suppose that $f \in (f)^2$; then $f = gf^2$ for some $g \in C([0, 1])$. Thus, for $x \neq \frac{1}{2}$, $g(x) = 1/f(x)$, and g is not continuous at $\frac{1}{2}$. We conclude that $f \notin (f)^2$.

Finally, (f) is not a prime ideal, since otherwise, by Problem 7.3, we would have $(f)^2 = (f)(f) = (f) \cap (f) = (f)$, which is a contradiction. Therefore, $(f) \neq M_{\frac{1}{2}}$, since $M_{\frac{1}{2}}$ is a maximal (and, hence, a prime) ideal.

Chapter 8

8.10   First, assume that R is a field. Let $f : R \to S$ be a homomorphism from R onto a nonzero ring S. Now, since R is a field, by Problem 2.20, the only ideals of R are (0) and R. But, $\ker(f) \neq R$ since f is onto and S is a nonzero ring; therefore, $\ker(f) = (0)$ and f is an isomorphism.

   Conversely, let I be a proper ideal of R, that is, $I \neq R$. Then the natural homomorphism $f : R \longrightarrow R/I$ is a homomorphism onto a non-zero ring R/I and is therefore, by hypothesis, an isomorphism. Thus, $I = \ker(f) = (0)$ and R has only two ideals, R itself and (0). We conclude, again by Problem 2.20, that R is a field.

Chapter 9

9.6   Since prime elements are always irreducible in integral domains, we only need show the converse is true for unique factorization domains. So, let $x \in D$ be irreducible and assume that $ab \in (x)$ for some $a, b \in D$. Then $ab = dx$ for some $d \in D$. But, D is a unique factorization domain, and so a and b have unique factorizations into products of irreducible elements. Clearly, then, by unique factorization, the irreducible element x, or an associate of x, must appear either in the factorization of a or in the factorization of b. Thus, $a \in (x)$ or $b \in (x)$ and x is prime.

9.11   Since x^2 and x^3 are both irreducible in $\mathbf{Z}[x^2, x^3]$, $\gcd(x^2, x^3) = 1$ and $\gcd(x^5, x^6) = x^3$. Two distinct factorizations for x^6 are

$$x^6 = x^2 \cdot x^2 \cdot x^2 = x^3 \cdot x^3.$$

Chapter 10

10.8   It is routine to check that the function $N(a + b\sqrt{5}) = a^2 - 5b^2$ is a norm function — that is,

$$N\big((a + b\sqrt{5})(c + d\sqrt{5})\big) = N(a + b\sqrt{5}) \cdot N(c + d\sqrt{5}).$$

The integer 4 has two factorizations into a product of irreducible elements as follows:

$$4 = 2 \cdot 2 = (3 + \sqrt{5})(3 - \sqrt{5}).$$

Each of 2, $3 + \sqrt{5}$, and $3 - \sqrt{5}$ are seen to be irreducible because they each have norm 4, and any proper divisor $a + b\sqrt{5}$ of any of them would have norm 2, which is impossible since then $a^2 - 5b^2 = 2$ implies that a^2 is ± 2 modulo 5, but only only 0 and ± 1 are squares modulo 5.

10.9   The integer 2 is irreducible — see Problem 10.8 — but it is not prime since $2 \mid (3 + \sqrt{5})(3 - \sqrt{5})$, and yet $2 \nmid (3 + \sqrt{5})$ and $2 \nmid (3 - \sqrt{5})$.

10.12   First, suppose that x is irreducible. Since D is a principal ideal domain, then x is also a prime element and (x) is a nonzero prime ideal. But, in principal ideal domains, nonzero prime ideals are maximal, so (x) is maximal.

Conversely, suppose (x) is maximal; then (x) is a prime ideal and x is a prime element. But, in any integral domain prime elements are irreducible, so x is irreducible.

Chapter 11

11.7   The maximal ideals are the principal ideals generated by the irreducible polynomials over K.

11.8   The prime ideals are (0), and then also all the maximal ideals, which are principal ideals generated by the irreducible polynomials over K.

11.10   First, assume $P[x]$ is prime in $R[x]$ and let $ab \in P$ for some $a, b \in R$. But, $ab \in P[x]$ also, and so, $a \in P[x]$ or $b \in P[x]$; hence, $a \in P$ or $b \in P$ and P is prime in R.

Conversely, suppose P is prime in R and let $fg \in P[x]$ for some $f, g \in R[x]$. Now, assume, by way of contradiction, that $f \notin P[x]$ and $g \notin P[x]$. Write $f = a_0 + a_1 x + a_2 x^2 + \cdots + a_n x^n$, and $g = b_0 + b_1 x + \cdots + b_m x^m$ and let a_i be the first coefficient of f which is not in P and let b_j be the first coefficient of g which is not in P. Then, the coefficient of x^{i+j} in the polynomial fg looks like

$$\cdots + a_{i-1}b_{j+1} + a_i b_j + a_{i+1}b_{j-1} + \cdots .$$

But all of the terms to the left of $a_i b_j$ in this coefficient are in P because of the choice of a_i and, similarly, all of the terms to the right of $a_i b_j$ are also in P because of the choice of b_j. Thus, $a_i b_j \in P$ which is impossible since P is prime. This contradiction proves that $P[x]$ is prime after all.

Alternatively, one could use Theorem 3.1 together with Problem 3.14.

11.11   Write $f = a_0 + a_1 x + a_2 x^2 + \cdots + a_n x^n$. First, assume all of the coefficients of f are nilpotent in R. But then, clearly, each of the individual elements $a_0, a_1 x, a_2 x^2, \ldots, a_n x^n$ are nilpotent in $R[x]$. Since the sum of nilpotent elements is also nilpotent, it follows that f is nilpotent in $R[x]$.

Conversely, assume f is nilpotent in $R[x]$. Then, by Theorem 5.2, f is in every prime ideal in $R[x]$. In particular, then, $f \in P[x]$ for all prime ideals $P[x]$ where P is a prime ideal in R (see Problem 11.10); that is, for all primes P in R, $f \in P[x]$, and so, for all primes P in R, all the coefficients of f are in P. Thus, by Theorem 5.2, all the coefficients of f are in the nilradical of R; that is, they are nilpotent in R.

11.16   By Theorem 3.1 the ideal $(x_1, \ldots, x_i)$ will be prime as long as the quotient ring $K[x_1, \ldots, x_n]/(x_1, \ldots, x_i)$ is an integral domain. But

$$K[x_1, \ldots, x_n]/(x_1, \ldots, x_i) \cong K[x_{i+1}, \ldots, x_n]$$

and, by Problem 3.15, $K[x_{i+1}, \ldots, x_n]$ is an integral domain. Thus, $(x_1, \ldots, x_i)$ is a prime ideal.

Chapter 12

12.7   Let $f, g \in I + (x)$, where $f = a_0 + a_1 x + \cdots$ and $g = b_0 + b_1 x + \cdots$. Then, $a_0, b_0 \in I$, but I is an ideal, so $a_0 - b_0 \in I$. Therefore, $f - g \in I + (x)$.

Next, let $f \in I + (x)$, where $f = a_0 + a_1 x + \cdots$; and let $g \in R[[x]]$, where $g = b_0 + b_1 x + \cdots$. Then, $a_0 \in I$, but again I is an ideal, so $a_0 b_0 \in I$. Therefore, $fg \in I + (x)$, and $I + (x)$ is an ideal.

12.11   Assume that $f \in R[x]$ is nilpotent in $R[x]$. Write $f = a_0 + a_1 x + \cdots$. Then, $f^n = 0$ for some n, and so $a_0{}^n = 0$ also and a_0 is nilpotent in $R[x]$. Thus, $f - a_0 = a_1 x + a_2 x^2 + \cdots$ is nilpotent in $R[x]$. But then, $a_1 x$ must be nilpotent in $R[x]$. Continuing in this way, each term $a_i x^i$ is nilpotent in $R[x]$. It of course follows that each coefficient a_i is nilpotent in R.

12.13   By Theorem 12.4, $M + (x)$ is a maximal ideal. Clearly, $M[[x]] \subset M + (x)$ since, for example, $x \in M + (x)$, but $x \notin M[[x]]$. Thus, $M[[x]]$ is not a maximal ideal.

Chapter 13

13.1   Let $I = (a_1, \ldots, a_n)$. We claim that $f(I) = \big(f(a_1), \ldots, f(a_n)\big)$. Let $b \in f(I)$; then $b = f(a)$ for some $a \in I$. Write $a = r_1 a_1 + \cdots + r_n a_n$; then

$$b = f(a) = f(r_1 a_1 + \cdots + r_n a_n) = f(r_1) f(a_1) + \cdots + f(r_n) f(a_n).$$

Thus, $b \in \big(f(a_1), \ldots, f(a_n)\big)$ and $f(I) \subseteq \big(f(a_1), \ldots, f(a_n)\big)$.

Now, let $s \in \big(f(a_1), \ldots, f(a_n)\big)$ and write $s = s_1 f(a_1) + \cdots + s_n f(a_n)$. But, since f is an onto homomorphism we can write each s_i as $f(r_i)$ for some $r_i \in R$. Thus, $s \in f(I)$ and $\big(f(a_1), \ldots, f(a_n)\big) \subseteq f(I)$. So, $f(I) = \big(f(a_1), \ldots, f(a_n)\big)$, as claimed.

13.14   We know from our discussion of localization in Chapter 6 that for any ideal J of R_T there is an ideal I of R such that $I_T = J$. Since R is Noetherian, I must be finitely generated, and so J is also finitely generated and R_T is Noetherian.

Chapter 14

14.1   Since R is an integral domain, (0) is a prime ideal (see Problem 3.2), and since $\dim R = 0$, (0) must also be a maximal ideal. Thus, R is a field by Problem 2.20, since the only ideals of R are (0) and R itself.

14.2   Since R is an integral domain, (0) is a prime ideal, and since R is not a field, R contains a maximal ideal $M \neq (0)$. Thus, $(0) \subset M$ is a chain of distinct prime ideals of length 1. But there can be no longer chain because, as we saw in Chapter 10, in a principal ideal domain every nonzero prime ideal is a maximal ideal. (In other words, if P and Q are two nonzero prime ideals of R such that $(0) \subset P \subseteq Q$ is a chain of prime ideals in R, then, since P is maximal, it follows that $P = Q$.)

14.6   Let

$$Q_0 \subset Q_1 \subset Q_2 \subset \cdots \subset Q_m$$

be a chain of distinct prime ideals in $R[x]$. Then, by Theorem 14.1,

$$Q_0 \cap R \subset Q_2 \cap R \subset Q_4 \cap R \subset \cdots$$

is a chain of distinct prime ideals in R. Therefore, $\lfloor \frac{m}{2} \rfloor \leq k$, from which it follows that $m \leq 2k + 1$. Thus, $\dim R[x] \leq 2k + 1$, as desired.

To verify the lower bound, let

$$P_0 \subset P_1 \subset P_2 \subset \cdots \subset P_k$$

be a chain of distinct prime ideals in R. Then,

$$P_0[x] \subset P_1[x] \subset P_2[x] \subset \cdots \subset P_k[x] \subset P_k[x] + (x)$$

is a chain of distinct prime ideals in $R[x]$ having length $k + 1$. Therefore, $\dim R[x] \geq k + 1$, as desired.

14.7   Let $a, b \in \sqrt{I}$ and $r \in R$. Then, $a^m \in I$ and $b^n \in I$ for some $m, n > 0$. But then $(ra)^m = r^m(a^m) \in I$, and so, $ra \in \sqrt{I}$. And, also, $(a - b)^{m+n} \in I$, and so, $a - b \in \sqrt{I}$. Therefore, $\sqrt{I}$ is an ideal.

14.8   Let $a \in \sqrt{I}$ and let P be a prime ideal containing I. Then, $a^m \in I \subseteq P$, for some $m > 0$, and so, $a \in P$.

Conversely, assume $a \notin \sqrt{I}$, then no power of a is in I. Thus, the set $S = \{1, a, a^2, a^3, \dots\}$ is disjoint from I. Now, using Zorn's lemma, among the ideals containing I, but disjoint from S, let P be a maximal element. We claim P is a prime ideal.

Suppose, then, $r \notin P$ and $s \notin P$. So, by the maximality of P, both (P, r) and (P, s) intersect S. Thus, for some $p, q \in P$, we have $p + xr = a^m$ and $q + ys = a^n$, which means that the product $(p + xr)(q + ys) = a^{m+n}$ is also in S. Now, if $rs \in P$, then this same product is also in P, which would be a contradiction, since P is disjoint from S. Therefore, $rs \notin P$, and P is prime, as claimed.

So, in conclusion, when $a \notin \sqrt{I}$, we produced a prime ideal P containing I such that $a \notin P$. This completes the proof.

14.10   Since $a \in M$ for each maximal ideal M, $1 + a \notin M$ for any maximal ideal M. Thus, $1 + a$ is not in the union of the maximal ideals of R, and so, by Theorem 5.1, $1 + a$ is a unit of R.

14.11   Let $I = (i_1, \dots, i_n)$, where n is the *least* positive integer such that I is generated by n nonzero elements. Since $I = JI$, we can write

$$i_1 = j_1 i_1 + \cdots + j_n i_n,$$

for some $j_1, \dots, j_n \in J$. So, we get

$$(1 - j_1)i_1 = j_2 i_2 + \cdots + j_n i_n.$$

But, by Problem 14.10, $1 - j_1$ is a unit, so, multiplying through by the inverse of $1 - j_1$, we see that $i_1 \in (i_2, \dots, i_n)$. Therefore, $I = (i_2, \dots, i_n)$, contradicting the minimality of n. Thus, $I = (0)$.

14.13   We use induction on k. For $k = 1$, the statement is trivial. For $k > 1$, assume, by way of contradiction, that I is not contained in any Q_i. Now, by the induction assumption, we also know that I is not contained in the union of any subcollection of the Q_is. So, for each i, we can let $a_i \in I$ where $a_i \in Q_i$, but $a_i \notin Q_j$ for all $j \neq i$.

Let $a = a_1 + a_2 a_3 \cdots a_k$. Then, $a \notin Q_1$, since $a_2 a_3 \cdots a_k \notin Q_1$. Similarly, for $i > 1$, $a \notin Q_i$, since otherwise, $a_2 a_3 \cdots a_k \in Q_i$ would imply that $a_1 \in Q_i$. Therefore, $a \in I$, but $a \notin Q_1 \cup \cdots \cup Q_k$. This contradiction completes the proof.

14.16   Let I be a non-zero ideal of $K[[x]]$. Now, among the power series in I, pick an f
         such that the *first* non-zero term of f has the least power n possible. Write
         $f = a_n x^n + a_{n+1} x^{n+1} + \cdots$. We claim that $I = (x^n)$. But, this is clear, because
         $f = x^n g$, where g is a unit by Theorem 12.1.

Chapter 15

15.3   In $K[x, y, z]$, we have $y^3 <_{\text{deglex}} xyz$, but $xyz <_{\text{degrevlex}} y^3$.

15.6   Assume $x_1^{a_1} x_2^{a_2} \cdots x_n^{a_n} | x_1^{b_1} x_2^{b_2} \cdots x_n^{b_n}$ and also, without loss, assume
         $x_1^{a_1} x_2^{a_2} \cdots x_n^{a_n} \neq x_1^{b_1} x_2^{b_2} \cdots x_n^{b_n}$. Then, $a_i \leq b_i$ for all i and, since the two
         monomials are not equal, $a_j \neq b_j$ for some j.
            For the lexicographic order this means that
         $x_1^{a_1} x_2^{a_2} \cdots x_n^{a_n} <_{\text{lex}} x_1^{b_1} x_2^{b_2} \cdots x_n^{b_n}$ since $a_j < b_j$ for the first j from the left
         where $a_j \neq b_j$.
            Now, by assumption, $a_1 + a_2 + \cdots + a_n < b_1 + b_2 + \cdots + b_n$, so both the
         degree lexicographic order and the degree reverse lexicographic order
         immediately give us $x_1^{a_1} x_2^{a_2} \cdots x_n^{a_n} <_{\text{deglex}} x_1^{b_1} x_2^{b_2} \cdots x_n^{b_n}$ and
         $x_1^{a_1} x_2^{a_2} \cdots x_n^{a_n} <_{\text{degrevlex}} x_1^{b_1} x_2^{b_2} \cdots x_n^{b_n}$ based solely upon the degrees of
         these two monomials without the need to resort to the tie-breaker.

15.7   Assume $x_1^{a_1} x_2^{a_2} \cdots x_n^{a_n} | x_1^{b_1} x_2^{b_2} \cdots x_n^{b_n}$, and also, without loss, assume
         $x_1^{a_1} x_2^{a_2} \cdots x_n^{a_n} \neq x_1^{b_1} x_2^{b_2} \cdots x_n^{b_n}$. Then, we have $x_1^{b_1} x_2^{b_2} \cdots x_n^{b_n} =$
         $(x_1^{a_1} x_2^{a_2} \cdots x_n^{a_n})(x_1^{c_1} x_2^{c_2} \cdots x_n^{c_n})$ where $c_i = b_i - a_i$ for each i.
            So, since $c_i \neq 0$ for some i by the assumptions, we have by the definition
         of a monomial order that

$$1 < x_1^{c_1} x_2^{c_2} \cdots x_n^{c_n},$$

         from which we can conclude, also by the definition of a monomial order, that

$$x_1^{a_1} x_2^{a_2} \cdots x_n^{a_n} < (x_1^{a_1} x_2^{a_2} \cdots x_n^{a_n})(x_1^{c_1} x_2^{c_2} \cdots x_n^{c_n})$$
$$= x_1^{b_1} x_2^{b_2} \cdots x_n^{b_n},$$

         as desired.

15.8   Suppose that for one of the monomials, $x_1^{i_1} x_2^{i_2} \cdots x_n^{i_n}$, the subset containment
         of ideals in the chain is not strict. This means that $x_1^{i_1} x_2^{i_2} \cdots x_n^{i_n}$ is an element
         of the ideal generated by the preceding monomials in the chain of monomials.
         We write this as

$$x_1^{i_1} x_2^{i_2} \cdots x_n^{i_n} \in \left(x_1^{a_1} x_2^{a_2} \cdots x_n^{a_n}, \ldots, x_1^{i_1} x_2^{i_2} \cdots x_n^{i_n} \right).$$

         Then $x_1^{i_1} x_2^{i_2} \cdots x_n^{i_n}$ can be expressed in terms of the generators of this ideal, so
         we can write

$$x_1^{i_1} x_2^{i_2} \cdots x_n^{i_n} = f_1 x_1^{a_1} x_2^{a_2} \cdots x_n^{a_n} + \cdots + f_i x_1^{i_1} x_2^{i_2} \cdots x_n^{i_n},$$

         where $f_1, \ldots, f_i \in K[x_1, x_2, \ldots, x_n]$. So, $x_1^{i_1} x_2^{i_2} \cdots x_n^{i_n}$ must appear somewhere
         as a product in this expression which means that for some preceding
         monomial, call it $x_1^{d_1} x_2^{d_2} \cdots x_n^{d_n}$, that monomial divides $x_1^{i_1} x_2^{i_2} \cdots x_n^{i_n}$. But

then it follows that $x_1{}^{d_1} x_2{}^{d_2} \cdots x_n{}^{d_n} < x_1{}^{j_1} x_2{}^{j_2} \cdots x_n{}^{j_n}$. This contradicts the fact that the chain of monomials is decreasing, and completes the proof.

15.9 If the total order is a well-ordering, then clearly the elements of the chain form a subset of S, and therefore have a least element which can then serve as the a_i where the chain terminates. Conversely, assume there is a subset of S that does not have a least element, then we easily construct, one element at a time, a strictly decreasing chain of elements from this subset that does not terminate, contrary to hypothesis.

15.10 We construct a set T just as if we were actually trying physically to divide a length a with a ruler of length b. We start with the length a and with our "ruler" measure off to a new length $a - b$, and then again to a length $a - 2b$, and so on, until the length is too short to measure again, and we have our remainder. With this physical motivation now in mind, we let

$$T = \{a - nb \mid n \in \mathbf{Z} \text{ and } a - nb \geq 0\}.$$

First, we show that T is nonempty. If $a \geq 0$, then $a \in T$ (for $n = 0$). On the other hand, if $a < 0$, then $a - ab \in T$, since $a - ab = a(1 - b)$ and $b \geq 1$. So, by the well-ordering principle, T has a least element r. Write $r = a - qb$.
Since, $r \in T, r \geq 0$. We must show that $r < b$. Suppose not, then $r = b + t$ where $t \geq 0$. But, then, $t = r - b = a - qb - b = a - (q + 1)b \in T$, and t is a smaller element than r in T, a contradiction. Thus, $r < b$, as desired.
We now prove the uniqueness of r and q. Assume, by way of contradiction, that $a = qb + r = pb + s$, where, without loss, $0 \leq r < s < b$. Then, $s - r = (a - pb) - (a - qb) = qb - pb = (q - p)b$, which is a contradiction, since $0 < s - r < b$. This completes the proof.

15.11 Assume, by way of contradiction, that $a = qg + r = pg + s$ where $r \neq s$, but $\deg(r) < \deg(g)$ and $\deg(s) < \deg(g)$, or, possibly, one of r and s is 0. Thus, $\deg(s - r) < \deg(g)$. But, $s - r = (a - pg) - (a - qg) = qg - pg = (q - p)g$, which means that $\deg(s - r) \geq \deg(g)$, a contradiction. This completes the proof.

15.12 Dividing $x^3 - x^2 + x - 1$ by $x^2 + x - 2$ we get a quotient $x - 2$ and a remainder $5x - 5$. We write $x^3 - x^2 + x - 1 = (x - 2)(x^2 + x - 2) + (5x - 5)$. Next, we divide $x^2 + x - 2$ by $5x - 5$, and we get a quotient $\frac{1}{5}x + \frac{2}{5}$ and a remainder 0. We write $x^2 + x - 2 = (\frac{1}{5}x + \frac{2}{5})(5x - 5) + 0$. Thus, $5x - 5$ is a greatest common divisor. Of course, then, $x - 1$ is also a greatest common divisor. Finally, working backward we can write

$$5x - 5 = 1 \cdot (x^3 - x^2 + x - 1) + (-x + 2)(x^2 + x - 2),$$

showing that $5x - 5$ is, in fact, in the ideal, and, hence, that it generates the ideal.

15.13 Beginning with $xy + 1$ yields a quotient y and a nonzero remainder $-x - y$. On the other hand, beginning with $y^2 - 1$ yields a quotient x and a remainder 0. Thus, $xy^2 - x \in (xy + 1, y^2 - 1)$.

15.14 The division algorithm stops before it even gets started here since x^3 is the leading term of both divisors and x^3 does not divide any term in $x^2 + x$. In particular, the division algorithm, even in $K[x]$, does not always produce a remainder of 0 when it should! Now, it is easy to see that $x \in (x^3, x^3 - x)$, because $x = x^3 - (x^3 - x)$. But then, we can also write $x^2 + x = (x + 1)x = (x + 1)(x^3 - (x^3 - x)) = (x + 1)x^3 + (-x - 1)(x^3 - x)$, as desired.

15.15   Clearly, if $x_1{}^{a_1} x_2{}^{a_2} \cdots x_n{}^{a_n}$ is divisible by one of the generating monomials of I, then $x_1{}^{a_1} x_2{}^{a_2} \cdots x_n{}^{a_n} \in I$. So, conversely, assume $x_1{}^{a_1} x_2{}^{a_2} \cdots x_n{}^{a_n} \in I$, and write

$$x_1{}^{a_1} x_2{}^{a_2} \cdots x_n{}^{a_n} = f_1 \cdot x_1{}^{b_{11}} x_2{}^{b_{12}} \cdots x_n{}^{b_{1n}} + \cdots + f_k \cdot x_1{}^{b_{k1}} x_2{}^{b_{k2}} \cdots x_n{}^{b_{kn}},$$

where $f_1, \ldots, f_k \in K[x_1, x_2, \ldots, x_n]$, and where the k monomials

$$x_1{}^{b_{11}} x_2{}^{b_{12}} \cdots x_n{}^{b_{1n}}, \ldots, x_1{}^{b_{k1}} x_2{}^{b_{k2}} \cdots x_n{}^{b_{kn}}$$

are generating monomials of I. Now, when the right-hand side of this expression is expanded, each term is divisible by one of the monomials $x_1{}^{b_{11}} x_2{}^{b_{12}} \cdots x_n{}^{b_{1n}}, \ldots, x_1{}^{b_{k1}} x_2{}^{b_{k2}} \cdots x_n{}^{b_{kn}}$. But, of course, the term $x_1{}^{a_1} x_2{}^{a_2} \cdots x_n{}^{a_n}$ must itself appear somewhere on the right-hand side, so it too must be divisible by one of the generating monomials

$$x_1{}^{b_{11}} x_2{}^{b_{12}} \cdots x_n{}^{b_{1n}}, \ldots, x_1{}^{b_{k1}} x_2{}^{b_{k2}} \cdots x_n{}^{b_{kn}}.$$

15.16   First, $z - x = z(xy + 1) - x(yz + 1) \in I$. The set $\{xy + 1, yz + 1\}$ is not a Gröbner basis for I for any of these monomial orders because in none of them do either of the leading terms xy or yz divide z or x.

15.20   Without loss, let $i_s = 1$ and $i_t = 2$. Let

$$g_1 = cx_1{}^{c_1} x_2{}^{c_2} \cdots x_n{}^{c_n} + \cdots \text{ and } g_2 = dx_1{}^{d_1} x_2{}^{d_2} \cdots x_n{}^{d_n} + \cdots$$

have leading terms $cx_1{}^{c_1} x_2{}^{c_2} \cdots x_n{}^{c_n}$ and $dx_1{}^{d_1} x_2{}^{d_2} \cdots x_n{}^{d_n}$, respectively. Let $e_i = \max(c_i, d_i)$ for each i — so that $x_1{}^{e_1} x_2{}^{e_2} \cdots x_n{}^{e_n}$ is the least common multiple of $x_1{}^{c_1} x_2{}^{c_2} \cdots x_n{}^{c_n}$ and $x_1{}^{d_1} x_2{}^{d_2} \cdots x_n{}^{d_n}$.
     Then,

$$S(x_1{}^{b_{11}} x_2{}^{b_{12}} \cdots x_n{}^{b_{1n}} g_1, x_1{}^{b_{21}} x_2{}^{b_{22}} \cdots x_n{}^{b_{2n}} g_2)$$

$$= \frac{1}{c} x_1{}^{b_{11}} x_2{}^{b_{12}} \cdots x_n{}^{b_{1n}} g_1 - \frac{1}{d} x_1{}^{b_{21}} x_2{}^{b_{22}} \cdots x_n{}^{b_{2n}} g_2$$

$$= \frac{x_1{}^{a_1} x_2{}^{a_2} \cdots x_n{}^{a_n}}{c\, x_1{}^{c_1} x_2{}^{c_2} \cdots x_n{}^{c_n}} g_1 - \frac{x_1{}^{a_1} x_2{}^{a_2} \cdots x_n{}^{a_n}}{d\, x_1{}^{d_1} x_2{}^{d_2} \cdots x_n{}^{d_n}} g_2$$

$$= \frac{x_1{}^{a_1} x_2{}^{a_2} \cdots x_n{}^{a_n}}{x_1{}^{e_1} x_2{}^{e_2} \cdots x_n{}^{e_n}} \left(\frac{x_1{}^{e_1} x_2{}^{e_2} \cdots x_n{}^{e_n}}{c\, x_1{}^{c_1} x_2{}^{c_2} \cdots x_n{}^{c_n}} g_1 - \frac{x_1{}^{e_1} x_2{}^{e_2} \cdots x_n{}^{e_n}}{d\, x_1{}^{d_1} x_2{}^{d_2} \cdots x_n{}^{d_n}} g_2 \right)$$

$$= \frac{x_1{}^{a_1} x_2{}^{a_2} \cdots x_n{}^{a_n}}{x_1{}^{e_1} x_2{}^{e_2} \cdots x_n{}^{e_n}} S(g_1, g_2).$$

Hence, if the remainder is 0 when the S-polynomial $S(g_1, g_2)$ is divided by the polynomials $g_1, \ldots, g_k$, then the remainder is also 0 for the S-polynomial $S(x_1{}^{b_{11}} x_2{}^{b_{12}} \cdots x_n{}^{b_{1n}} g_1, x_1{}^{b_{21}} x_2{}^{b_{22}} \cdots x_n{}^{b_{2n}} g_2)$.

15.21   Dividing first by $x + z$ we get a quotient z. Multiplying and subtracting
produces a new polynomial $yz - z^2$. Dividing now by $y - z$ gives us once again a
quotient z, and the next multiplication and subtraction results in a remainder
0. (Of course, this also allows us to write $xz + yz = z(x + z) + z(y - z)$.)

15.22   There are three S-polynomials to check. We first check

$$S(x^2, y^3) = \frac{x^2 y^3}{x^2} x^2 - \frac{x^2 y^3}{y^3} y^3 = x^2 y^3 - x^2 y^3 = 0,$$

and the S-polynomial itself is 0.
   Next, we check

$$S(xy + y^2, y^3) = \frac{xy^3}{xy}(xy + y^2) - \frac{xy^3}{y^3} y^3 = xy^3 + y^4 - xy^3 = y^4,$$

and, since this S-polynomial y^4 is clearly divisible by y^3, it also has remainder 0.
   Finally, we check

$$S(x^2, xy + y^2) = \frac{x^2 y}{x^2} x^2 - \frac{x^2 y}{xy}(xy + y^2) = x^2 y - x^2 y - xy^2 = -xy^2.$$

Dividing $-xy^2$ first by $xy + y^2$, we get a quotient $-y$. Multiplying and
subtracting produces a new polynomial y^3, which we can now divide by y^3 to
yield a remainder 0.
   Since all three S-polynomials yielded a remainder 0, the set $\{x^2, xy + y^2, y^3\}$
is a Gröbner basis, by Theorem 15.7.

15.23   For $x > y > z$ we first write each of these polynomials in descending order, and
then we get

$$S(-x^2 + y, -x^3 + z) = \frac{x^3}{-x^2}(-x^2 + y) - \frac{x^3}{-x^3}(-x^3 + z)$$

$$= x^3 - xy - x^3 + z = -xy + z.$$

From this we see that the set $\{y - x^2, z - x^3\}$ is not a Gröbner basis for I because
neither of the leading terms, $-x^2$ or $-x^3$, divides either term of the
S-polynomial $-xy + z$.
   On the other hand, for $z > y > x$ we get

$$S(y - x^2, z - x^3) = \frac{zy}{y}(y - x^2) - \frac{zy}{z}(z - x^3)$$

$$= zy - zx^2 - zy + yx^3 = -zx^2 + yx^3.$$

Then, dividing $-zx^2 + yx^3$ first by z — the leading term of $z - x^3$ — gives us a
quotient $-x^2$. Multiplying and subtracting produces a new polynomial
$yx^3 - x^5$. Dividing now by $y - x^2$ yields a quotient x^3, and the next
multiplication and subtraction results in a remainder 0. Therefore, with this
monomial order, the set $\{y - x^2, z - x^3\}$ is a Gröbner basis for I.

Suggestions for Further Reading

Problem 1. (on page 211) First, assume that the Krull dimension of R is at most n. So, we can let dim $R = k$ where $k \leq n$ and, further, we can let

$$P_0 \subset P_1 \subset P_2 \subset \cdots \subset P_k$$

be a chain of distinct prime ideals of R having having that maximum possible length k. In particular, then, P_k is a maximal ideal (otherwise, we could increase the length of the chain).

Next, we pick $x \in R$ and we must show that the Krull dimension of the boundary R_T is at most $k - 1$. We claim that the maximal ideal P_k is not disjoint from the multiplicatively closed set T. Of course, this is clear if $x \in P_k$ (since we can write $x = x^1(1 + 0x) \in T$). On the other hand, if $x \notin P_k$, then $(P_k, x) = R$ by the maximality of P_k, and so, $1 = m + rx$ for some $m \in P_k, r \in R$. So, we can write $m = 1 - rx$, and then, $xm = x(1 + (-r)x) \in T \cap P_k$. Thus, as claimed, P_k is not disjoint from T.

Now, recall that, by Theorem 6.1, there is an order preserving correspondence between the prime ideals of the localization R_T and the prime ideals of R which are *disjoint* from T. Let us suppose, by way of contradiction, that R_T contains a chain of prime ideals of length k. This chain in R_T would then, by Theorem 6.1, correspond to a chain $P_0 \subset P_1 \subset P_2 \subset \cdots \subset P_k$ of prime ideals in R as above in which the primes are all *disjoint* from T. But we have seen that P_k is not disjoint from T, which is a contradiction. Thus, the Krull dimension of R_T is at most $k - 1$, as desired.

Conversely, assume that for each $x \in R$ the Krull dimension of the boundary R_T is at most $n - 1$. Again, by way of contradiction, assume that R has a chain of prime ideals of length greater than n and let

$$P_0 \subset P_1 \subset P_2 \subset \cdots \subset P_{n+1}$$

be such a chain where P_{n+1} is a maximal ideal. (We are not necessarily assuming, by the way, that dim $R = n + 1$.)

Next, we pick an $x \in P_{n+1} \backslash P_n$, and we claim that for this x all of the prime ideals $P_0, P_1, P_2, \ldots, P_n$ will be disjoint from the multiplicatively closed set T. Clearly, it is sufficient to verify that P_n and T are disjoint. So, suppose that some element $x^i(1 + rx)$ in T is *also* an element of P_n. But, $x^i(1 + rx) \in P_n$, where $x \notin P_n$ and P_n is a prime ideal, which means that $1 + rx \in P_n$. This, in turn, means that $1 + rx \in P_{n+1}$, which is impossible, since $x \in P_{n+1}$, and this would imply that $1 \in P_{n+1}$, contradicting the fact that P_{n+1} is a maximal ideal of R. Thus, for this choice of x all of the prime ideals $P_0, P_1, P_2, \ldots, P_n$ are disjoint from T.

Now, by Theorem 6.1, the chain

$$P_0 \subset P_1 \subset P_2 \subset \cdots \subset P_n$$

of prime ideals in R — since they are all disjoint from T — corresponds to a chain of prime ideals in R_T of length n. But, this is a contradiction since the Krull dimension of the boundary R_T is at most $n - 1$. This contradiction shows that, contrary to our original assumption, R cannot have a chain of prime ideals of length greater than n. Thus, the Krull dimension of R is at most n. This completes the proof.

Suggestions for Further Reading

Algorithms such as Buchberger's algorithm come to a natural end, but books — and journeys — do not. You just have to stop somewhere. Fortunately, there are many excellent books and articles that can serve as guides for further study. Here is an abbreviated list of some I recommend.

There are two classics I find especially helpful. The Atiyah and Macdonald book is a thin concise volume that contains a wealth of material, and Kaplansky's book undoubtedly has more beautiful mathematical proofs than any other book I know. Many of these managed to find their way into the text you are now holding.

M. F. Atiyah and I. G. Macdonald, *Introduction to Commutative Algebra*, Addison-Wesley, 1969; paperback, 1994.

Irving Kaplansky, *Commutative Rings*, revised edition, University of Chicago Press, 1974; Polygonal Publishing House, 1994.

For a general text on the subject of commutative rings, the one by Matsumura can't be beat.

Hideyuki Matsumura, *Commutative Ring Theory*, second edition, Cambridge University Press, 1989.

The notoriously difficult path into the study of algebraic geometry has been made much easier, and far more enjoyable, in recent years due to the following book that, not only is wonderfully written, but has been written especially with undergraduates in mind:

David Cox, John Little, and Donal O'Shea, *Ideals, Varieties and Algorithms: An Introduction to Computational Algebraic Geometry and Commutative Algebra*, second edition, Springer-Verlag, 2006.

For an advanced and extremely thorough treatment of commutative algebra, and one that always keeps geometry firmly in the picture, I would suggest trying

David Eisenbud, *Commutative Algebra: with a View Toward Algebraic Geometry*, Springer-Verlag, 1995.

Another extraordinarily good book, and again one that has been written with undergraduates in mind, is one that takes a particularly algorithmic point of view:

William W. Adams and Philippe Loustaunau, *An Introduction to Gröbner Bases*, American Mathematical Society, 1994.

There is a marvelous two-volume encyclopedic set of books on what is, after all, the relatively new area of "computational" commutative algebra. These books are chock full of theory, examples, exercises, applications, and even tutorials using the computer algebra system CoCoA.

Martin Kreuzer and Lorenzo Robbiano, *Computational Commutative Algebra 1*, Springer-Verlag, 2000.

Martin Kreuzer and Lorenzo Robbiano, *Computational Commutative Algebra 2*, Springer-Verlag, 2005.

For example, in the Introduction of Volume 1 you can discover a truly amazing fact about polynomials. There exists a polymomial whose square has *fewer* terms than the polynomial itself! This is not a question that Fermat or Euler could have resolved in his day. But, by page 262, Kreuzer and Robbiano have shown the reader how to produce a polynomial

$$f = x^{12} + \frac{2}{5}x^{11} - \frac{2}{25}x^{10} + \frac{4}{125}x^9 - \frac{2}{125}x^8 + \frac{2}{125}x^7 - \frac{3}{2750}x^6$$

$$- \frac{1}{275}x^5 + \frac{1}{1375}x^4 - \frac{2}{6875}x^3 + \frac{1}{6875}x^2 - \frac{1}{6875}x - \frac{1}{13750},$$

which has 13 terms, and yet its square f^2 has only 12 terms.

I urge readers to experiment with some of the widely available computer algebra systems such as Maple, Mathematica, and Macsyma, or CoCoA and Macauley, which are both available free of charge and can be downloaded from:

ftp://cocoa.dima.unige.it/cocoa/index.html

and

http://www.math.uiuc.edu/Macaulay2/Downloads/index.html

Getting back now to my list of recommended readings, the book that formed the basis of Chapter 7 is quite simply one of the finest mathematics books ever written:

Leonard Gillman and Meyer Jerison, *Rings of Continuous Functions*, Van Nostrand Reinhold, 1960.

One of my all-time favorite books in commutative ring theory is a highly unusual one, written by a student of Kaplansky's by the way, that contains nothing but examples — 180 in all — of commutative rings with various properties:

Harry Hutchins, *Examples of Commutative Rings*, Polygonal Publishing House, 1981.

A very specialized but highly readable book pursues one particularly fascinating topic in great detail:

James W. Brewer, *Powers Series Rings over Commutative Rings*, Marcel Dekker, 1981.

A single paper, more than any other, influenced my perspective throughout the entire text, but especially in Chapters 11 and 12:

Robert Gilmer, On polynomial and power series rings over a commutative ring, *Rocky Mountain Journal of Mathematics* 5 (1975), 157-175.

Another important paper is an extremely rich one containing Theorems 14.3 and 14.4 that were used to prove that dim $K[x_1, \ldots, x_n] = n$, and to prove that for Noetherian rings dim $R[x_1, \ldots, x_n] = n + \dim R$. And, that's just the tip of the iceberg for this remarkable paper:

J. W. Brewer, W. J. Heinzer, P. R. Montgomery, and E. A. Rutter, Krull dimension of polynomial rings, *Proceedings: Kansas Conference on Commutative Algebra*, Springer-Verlag Lecture Notes in Mathematics 311 (1973), 26-45.

A recent paper in the *Monthly* also presents an 'elementary' proof that dim $K[x_1, \ldots, x_n] = n$:

Thierry Coquand and Henri Lombardi, A short proof for the Krull dimension of a polynomial ring, *American Mathematical Monthly* 112, November 2005, 826-829.

They begin by defining a localization, called the *boundary*:

Definition 1. *Let R be a ring and $x \in R$. Then, the* **boundary** *of x in R is the localization R_T, where T is the multiplicative system*

$$ T = \{x^n(1 + rx) \mid n \in \mathbf{N}, r \in R\}. $$

The reason for the term "boundary" is that, if x is a nilpotent element or if x is a unit, then $R_T = 0$. They then characterize Krull dimension in terms of this localization without any direct reference to chains of prime ideals.

Theorem 1. *Let R be a ring and let n be a non-negative integer. Then the Krull dimension of R is at most n if and only if for each $x \in R$ the Krull dimension of the boundary R_T is at most $n - 1$.*

For example, if we agree that for the zero ring $R = 0$ the Krull dimension is to be -1, then this theorem will correctly tell us that dim $K = 0$, for a field K. Here is one final exercise for you to do; obviously you'll need Theorem 6.1.

Problem 1. Prove Theorem 1.

But, getting back to the reading list: one of the best ways to study the history of a subject is to read letters by the mathematicians themselves:

G. Sabidussi, Correspondence between Sylvester, Petersen, Hilbert and Klein on invariants and the factorization of graphs 1889-1891, *Discrete Math.* 100 (1992), 99-155.

Another excellent historical source, not only on Fermat's Last Theorem, but on the overall historical development of modern algebra is

Harold M. Edwards, *Fermat's Last Theorem: A Genetic Introduction to Algebraic Number Theory*, Springer-Verlag, 1996.

And lastly, I strongly recommend that you go back to where it all began, and take a look at a wonderful translation by Reinhard Laubenbacher of a book based on a course that Hilbert taught in 1897 on invariant theory at the University of Göttingen:

David Hilbert, *Theory of Algebraic Invariants*, Cambridge University Press, 1993.

Index

(*a*), 14
[*a*], 53
(*a*, *b*), 21
$a + I$, 15
Ann(*a*), 197
annihilator ideals, 197
aR, 15
Arithmetica, 94–95
ascending chain condition, 104, 128
ascending chains, 38, 102, 128
associates, 89
Axiom of Choice, 41–43, 160

Bachet, translation of Diophantus, 94
Bézout domains, 108
boundary, 211
bounded support, 76
Brewer, Jim, 144
Buchberger, Bruno, 156, 172
Buchberger's algorithm, 176–77
Buchberger's theorem, 172–75

C, 4
$C([0,1])$, 70
Cantor, Georg, 42–43
Cauchy, Augustin-Louis, 97
$c(f)$, 113
chains, 37, 157; ascending, 38, 102, 128; descending, 38; of prime ideals, 118
closed subsets, 7, 11
CoCoA, 156, 209, 210
codimension, 140
Cohen, I. S., 130
Cohen, Paul, 43
commutative rings with identity, 6
compact topological spaces, 76, 79
constant terms, 13
content, of a polynomial, 113
continuum hypothesis, the, 43
cosets, 15; addition of, 17; basic properties of, 16; multiplication of, 17; representatives of, 15
$C(\mathbf{R})$, 22
$C(X)$, 76

Dedekind, Richard, 11, 14, 36, 97
degree, of a polynomial, 13
degree function, Δ, the, 104
degree lexicographic order, $<_{\text{deglex}}$, 158

degree reverse lexicographic order, $<_{\text{degrevlex}}$, 159
Δ, 104
dimension, 118; Krull, 137–51
dim *R*, 137
Diophantus, 94
Dirichlet, Adrien-Lejeune, 96
divide (*m*|*n*), definition of, 23, 89
division algorithm, the, 161; for $K[x]$, 161; for $K[x_1,\dots,x_n]$, 164
domains, 25; Bézout, 108; Euclidean, 104; Gaussian, 112; GCD, 108; integral, 25; principal ideal, 100; unique factorization, 92

equivalence classes, 53; basic properties of, 54
equivalence relations, $\cong$, 51
equivalent factorizations, 93
Euclid, 104, 107
Euclidean algorithm, the, 104, 107–8, 190
Euclidean domains, 104
Euler, Leonard, 95–96
Euler's formula, 119

Faltings, Gert, 95–96
Fermat, Pierre de, 93–95, 109; Last Theorem, 93–98, 211; method of infinite descent, 94, 96, 174; principle of least time, 94; two square theorem, 109
fields, 21, 25
Fields, D., 121, 127
Frey, Gerhard, 95
fundamental theorem of arithmetic, the, 90

Gauss, Carl Friedrich, 11
Gaussian integers, 49, 98, 105, 107
Gauss's lemma, 114
GCD-domains, 108
gcd(*x*, *y*), 99
Germain, Sophie, 95
Germain primes, 95
Gödel, Kurt, 43
Gordan, Paul, 1, 2, 156
Gordan's Problem, 1
greatest common divisor, 99, 108
Gröbner, Wolfgang, 156
Gröbner basis, 155, 166–79; minimal, 178; reduced, 155, 178